Cultures of Prediction

Engineering Studies

Edited by Gary Downey and Matthew Wisnioski

Matthew Wisnioski, *Engineers for Change: Competing Visions of Technology in 1960s America*

Amy Sue Bix, *Girls Coming to Tech! A History of American Engineering Education for Women*

Jessica Smith, *Extracting Accountability: Engineers and Corporate Social Responsibility*

Elizabeth Reddy, *¡Alerta! Engineering on Shaky Ground*

Konstantinos Chatzis, *Forecasting Travel in Urban America: The Socio-Technical Life of an Engineering Modeling World*

Ann Johnson and Johannes Lenhard, *Cultures of Prediction: How Engineering and Science Evolve with Mathematical Tools*

Cultures of Prediction

How Engineering and Science Evolve with Mathematical Tools

Ann Johnson and Johannes Lenhard

The MIT Press
Cambridge, Massachusetts
London, England

The MIT Press would like to thank the anonymous peer reviewers who provided comments on drafts of this book. The generous work of academic experts is essential for establishing the authority and quality of our publications. We acknowledge with gratitude the contributions of these otherwise uncredited readers.

This book was set in Stone Serif and Stone Sans by Westchester Publishing Services. Printed and bound in the United States of America.

Library of Congress Cataloging-in-Publication Data

Names: Johnson, Ann, 1965–2016, author. | Lenhard, Johannes, author.
Title: Cultures of prediction : how engineering and science evolve with mathematical tools / Ann Johnson, Johannes Lenhard.
Description: Cambridge, Massachusetts : The MIT Press, 2024. | Series: Engineering studies | Includes bibliographical references and index.
Identifiers: LCCN 2023036534 (print) | LCCN 2023036535 (ebook) | ISBN 9780262548236 (paperback) | ISBN 9780262379052 (epub) | ISBN 9780262379045 (pdf)
Subjects: LCSH: Mathematical models—History. | Engineering mathematics—History. | Predictive analytics. | Predictive control.
Classification: LCC TA342 .J64 2024 (print) | LCC TA342 (ebook) | DDC 620.001/51—dc23/eng/20240118
LC record available at https://lccn.loc.gov/2023036534
LC ebook record available at https://lccn.loc.gov/2023036535

10 9 8 7 6 5 4 3 2 1

Contents

Series Foreword

We live in highly engineered worlds. Engineers play crucial roles in the normative direction of localized knowledge and social orders. The *Engineering Studies Series* highlights the growing need to understand the situated commitments and practices of engineers and engineering. It asks: what is engineering for? What are engineers for?

Drawing from a diverse arena of research, teaching, and outreach, engineering studies raises awareness of how engineers imagine themselves in service to humanity, and how their service ideals impact the defining and solving of problems with multiple ends and variable consequences. It does so by examining relationships among technical and nontechnical dimensions, and how these relationships change over time and from place to place. Its researchers often are critical participants in the practices they study.

The *Engineering Studies Series* publishes research in historical, social, cultural, political, philosophical, rhetorical, and organizational studies of engineers and engineering, paying particular attention to normative directionality in engineering epistemologies, practices, identities, and outcomes. Areas of concern include engineering formation, engineering work, engineering design, equity in engineering (gender, racial, ethnic, class, geopolitical), and engineering service to society.

The *Engineering Studies Series* thus pursues three related missions: (1) advance understanding of engineers, engineering, and outcomes of engineering work; (2) help build and serve communities of researchers and learners in engineering studies; and (3) link scholarly work in engineering studies to broader discussions and debates about engineering education, research, practice, policy, and representation.

Gary Lee Downey and Matthew Wisnioski, Editors

Preface

The history of this book has had a significant impact on its structure and content. More than fifteen years ago, I met Ann Johnson at an interdisciplinary conference. As a historian of engineering, she had a strong interest in computational methods and an intellectual appetite spanning history, philosophy, and science studies. Although she was well aware that these fields are marked by well-attended disciplinary boundaries, she had the courage and the bounce to jump over them. We immediately realized that there was a sort of common wavelength that made our ideas, perspectives, praises, and criticisms coincide to an extent that surprised us both. That this book attempts to sit comfortably on, and partly in between, the chairs of history of engineering, history of technology, philosophy of science, and social studies of science is dictated as much by its subject as by the way it fits both Ann's and my own preferences (the latter being deeply influenced by the progress of our joint work).

We developed the overall perspective of this book more than a decade ago. Starting from the observation that many accounts of the computer were geared toward fast and gigantic machines, we asked what—if anything—characterized the impact of easily available, relatively cheap networked computers. Discussing this question led us to the claim of "a new culture of prediction" (Johnson and Lenhard 2011) that soon amplified into a longue-durée perspective on mathematization detailed in the introduction to this book. However, neither of us felt any rush to finish this book swiftly. It was the joys of gathering material, cobreeding ideas, recombining plans, and looking for additional perspectives that dominated.

Ann's unexpected passing in 2016 terminated the project. Two years later, the newly founded Ann Johnson Institute revived it. My generously funded stay at Columbia, South Carolina, where Ann had been a professor before moving to Cornell, reactivated intellectual and atmospheric memories in a way that turned the writing of this book into a doable and, more importantly, an enjoyable enterprise. In early 2018, the only jointly discussed and revised text we had in stock was a chapter on the "epistemology of iteration" (now chapter 4). It examines the history of computational chemistry, spanning different cultures of prediction starting in precomputer times. Furthermore, Ann had worked on a chapter that investigates how software code moves (concentrating on fluid dynamics). As usual, she tested preliminary versions in workshops and conferences. This text forms chapter 6, and no doubt my well-intentioned additions represent only a taste of the fruitful ways in which Ann would have further researched and elaborated the text. Beyond these two chapters, I had to reinvent and reexperience our joint discussions—fortunately with the help of a treasure chest of sketches. To my delight, I found that Ann's voice is not silent and that her spirit is still active. Of course, Ann's forceful voice would have had a much stronger and beneficial impact than the echo in my imagination. Who knows what this book would look like if Ann were still alive? Whatever the case, her input can be found on all levels, from methods to outlines and written material, making it truly a coauthored book.

Acknowledgments

Ann and I owe a great debt of gratitude to a lot of people. Only through wise and determined nudging in the right direction has the project been able to maintain, and sometimes regain, momentum. And I would like to add that without the encouragement, stimulation, and criticism in response to provisional chapters, I would hardly have been able to realize our jointly planned and launched project. This happened in reading groups, colloquia, and conferences that I do not want to list here—and that I probably could not list. Translating my gratitude into a concrete list is almost a historical project, given that Ann's and my work on this book started well over a decade ago.

Location and personal contacts are of great importance, especially for an interdisciplinary project such as Ann and I were aiming for in the case of prediction. Longer-term visits and joint work were supported by the

Humboldt Foundation, DAAD, DFG (grant LE 1401/7–1), and, of course, the Ann Johnson Institute.

Different stages of the book have profited from the helpful advice provided by scholars in history, philosophy, and engineering departments located mainly, but not exclusively, in the University of South Carolina, Columbia, and Bielefeld University's Center for Interdisciplinary Research (ZiF) in Germany. These include Davis Baird, Martin Carrier, Hans Hasse, Ron Kline, Allison Marsh, Leah McClimans, Anne Marcovich, Cornelis Menke, Robert Mullen, Joe November, Michael Otte, Carsten Reinhardt, Terry Shinn, Michael Stöltzner, and Heidi Voskuhl. It was a pleasure to receive the constructive help of Matt Wisnioski and Gary Downey, the editors of the Engineering Studies series at the MIT Press, and also of the reviewers of the manuscript whose comments were very helpful when making revisions. A host of special thanks go to Matthias Brandl for his thoughtfulness, his feeling for language, and his willingness to share energy. The gap between my attempts to write in plain English and the actual text was narrowed in a beneficial way by Jonathan Harrow.

1 Introduction

This book is about predictions. Throughout history, the imagination of human societies has been captured by the thought of tearing a hole in the fog of uncertainty and looking into the future. Croesus consulted the Delphi oracle, the leading producer of predictions for policy advice at the time, and then went to war against the Persians. Unfortunately, this cost him his kingdom: he had misinterpreted the oracle's prediction that a great empire would perish. To give a more recent example, climate change is perceived as a grave global problem. National policies and international treaties are calibrated according to the ensuing change in temperature, with its prediction relying on computational models and observational data of breathtaking dimensions. In short, from the sacred to the mundane, from the long-term to the immediate, predictions guide actions, sometimes to bring about and sometimes to avoid what has been predicted—sometimes with success, sometimes with failure.[1]

With this in mind, we examine a certain subset of predictions: those made by scientists and engineers with the help of mathematical tools. Making predictions constitutes one of the dominant features of science. In fact, the ability to make reliable predictions based on robust and replicable methods is possibly the most distinctive and important advantage claimed for scientific knowledge compared to other types of knowledge. Engineers' claims of expert knowledge and its utility for the modern world stem from their special ability to predict whether a bridge or building will stand up or where a cannonball will strike. Although their abilities to predict are specific and not identical to those of science, they are closely connected. Since at least the seventeenth century, engineers have drawn on science, and especially mathematics, to make predictions about the human-built world that they design. Despite the obvious importance of prediction to modern engineering and scientific enterprise, it is a strangely understudied element. When prediction is studied, it is treated as a monolithic block—static in time and a fortunate outcome of the mathematization and mechanization of the world picture.

In this book, we reveal a very different picture by looking at the *practices* of making predictions in science and engineering. Studying these practices over four centuries exposes a dynamic picture of prediction-making and shows that mathematization is a flexible concept. Moreover, different modes of prediction, complementary concepts of mathematization, and technology *coevolved*, thereby building patterns that we call *cultures of prediction*.

This new orientation toward prediction challenges some of the basic tenets of the philosophy of science in which scientific theories and models are seen as predominantly explanatory rather than predictive.[2] Predictive strategies in general, and exploratory and iterative ones in particular, introduce a design mode of knowledge production that focuses on making things and testing their performance. In computer-based approaches such as simulation or machine learning, the value of predictive models increases. This makes cultures of prediction commonplace across a wide array of research areas in both engineering and science (though surely not universal). These cultures of prediction, in turn, inform the sort of research projects that scientists and engineers undertake and the kinds of endeavors that funding agencies support, thus reinforcing the turn toward prediction.

1.1 The Landscape

Largely due to the role of the computer, the phenomenon of making predictions in science and engineering is a hot topic today with a growing number of popular books that introduce their readers to new ways of predicting.[3] Academic titles on computers, data production, and predictions are also gaining momentum.[4] Although this book is an academic one, it also invites readers with a more general interest because the history of mathematization provides an illuminating background to the recent upswing in data- and computer-driven predictions. To improve readability, we have kept technical aspects to a minimum.

Two clusters of existing literature are particularly central for this book. The first examines *systems* in which computer technology forms a defining component.[5] Accounts such as Paul Edwards's *The Closed World* (1996) highlight how computer technology, methodologies for creating predictions, and institutional organization coevolve.[6] Venturing into complexity, building computer systems and a political (often also military) demand for prediction reinforce each other. Prediction in this context is close to control because the ability to control a system typically requires predicting how it will behave under certain conditions.[7]

The second cluster revolves around mathematization and contains a number of studies in the history as well as philosophy of science that attempt to explain how nature has been addressed mathematically to produce new knowledge. Most of these studies focus on either the period of the scientific revolution or the turn of the twentieth century, when developments in physics were central to the project of mathematization.[8] This study looks at these periods as well but does so over the longue-durée, from the sixteenth century to the present, and includes the move to mechanical and electronic calculation and computation.[9]

Our use of culture and practices owes much to recent accounts in the history, sociology, and philosophy of science and technology. The ensuing and sometimes cross-disciplinary debates are among the truly illuminating episodes in studies of science. For more than two decades now, a practice-oriented approach has gained traction.[10] We would like to mention three points around which this literature has informed our endeavor: First, there are highly valuable studies on how computing technology, institutions, infrastructure, and concepts such as certainty depend on one another.[11] Second, we learned about the conceptual difficulties of capturing interdependence and coevolution.[12] Coevolution is pivotal to our dynamic history of predictive methods and values in science and engineering. Third, it is important to examine engineering as a site of knowledge production because prediction stands among the highest epistemic goals of engineering work. It may well be that the cultures and values of prediction have been overlooked by scholars because the epistemology of engineering has itself been overshadowed by physics. We shift that frame here and focus primarily on prediction in engineering and chemistry and on fields such as statistics and machine learning for our examples. We believe that changing the frame in this way could also shift prediction in physics and biology toward a more central position.

Although our book unfolds amid strong bodies of historical, philosophical, and sociological literature, it investigates a space that has not yet been covered because we bring into contact and weave together several elements in an innovative way: first, the history and philosophy of prediction; second, the important changes that desktop and networked computing have brought to scientific endeavors; third, the history of engineering and applied science as paradigmatic cases with which to understand fundamental elements in the structure of scientific knowledge and epistemological practices in science more generally; and fourth, a historical study of mathematization outside of the history of physics. In our account here, mathematization undergirds the history of prediction.[13]

1.2 Four Cultures of Prediction

Cultures of prediction draw attention to four points: First, the notion of cultures stresses that mathematization, epistemology, technology, and social organization coevolve. Second, modes of mathematization and prediction refer to the status of mathematics and the ways in which it is used to create predictions. These modes develop rather slowly and take at least a generation to gain a foothold. Older modes and cultures persist while new ones develop alongside them.[14] Third, we do not claim that the modes and cultures apply to all of science and engineering. We emphasize the plurality of mathematical instrumentation and the related differences in predictive cultures. Fourth, this book takes a particular focus on today's culture of prediction and its use of computational tools. With these tools, computer technology contributes to (and coevolves with) practices of prediction in new ways. We take *a long-term perspective and examine the dynamic history of predictive methods and values in science and engineering. This enables us to gain a new perspective on today's uses of, and sometimes obsessions with, predictions.*[15]

We identify four different cultures of prediction in the history of science and engineering: rational, empirical, iterative–numerical, and exploratory–iterative. Factoring in these cultures of prediction leads to a revaluation of epistemology.

The *rational* and *empirical* cultures originated in the sixteenth and seventeenth centuries as part of the initial shaping of modern science. They developed in close and sometimes conflictual interrelation. In the rational culture, mathematics captures the structure of physical phenomena. Because the book of nature is written in mathematical symbols, as Galileo expressed so famously, and because general laws are important parts of the mathematical structure, one can derive predictions mathematically. The empirical culture does not so much dwell on the generality and rigidity of derivation but on the flexibility with which mathematical models can be adjusted to observations and experience. Whether a mathematical model will produce predictions does not depend on what grounds it has been construed on. Both cultures present something like ideal types that provide an orientation for practitioners, shape their expectations, and direct their energies. What we see in history is not a perfect realization of the rational or empirical culture but rather mixed, often tense and controversial, situations in which rational and empirical components compete for dominance and, typically for engineering, build *hybrids*.

The third, *iterative–numerical* culture, emerged with technologies of calculation in the nineteenth century. The iterative–numerical mode of prediction predated the computer but became fully established with mainframe machines. Predictions via iteration of simple computational algorithms became an option. From this new perspective on prediction, the iterative capacity of mathematical instruments rose to become an important factor. Of course, the iterative–numerical culture reached full bloom only after the advent of the digital electronic computer. The computer redirected the path of mathematization. At the same time, the big and expensive mainframe machines were institutionally organized in a peculiar way: in centralized systems.

We claim that the networked desktop computer gives rise to a fourth, *exploratory–iterative*, culture that is distinct from the iterative–numerical culture.[16] "Working on" models means exploring the relationships between input data and output data to produce predictions that are then often used to modify the models before the next stage of exploration. Such a practice calls for easy and fast feedback between models and modelers who modify parts and adapt parameters. In the previous era when computing meant waiting for time on a mainframe processor or even sending in punch cards for batch processing, "tweaking" models—that is, running versions of models with minor changes—was impractical. The exploratory nature of computational modeling therefore depends on cheap and convenient local access to computational power. This accessibility came only with the mature, networked desktop computer around the 1990s.

1.3 Overview of the Chapters

Chapter 2 discusses the entangled history of the rational and empirical cultures of prediction by looking at the history of ballistics. Ballistics is an exemplary case for prediction that highlights the polarity between rational and empirical in a long-term perspective spanning several centuries that throws light on numerous promises and frustrations. The chapter unfolds the case for prediction by examining one episode each for the sixteenth, seventeenth, and eighteenth centuries.

Chapter 3 focuses on mechanical engineering when mathematization took hold in the late nineteenth century and the rational–empirical distinction received new interpretations. The role of mathematics was hotly debated and deeply controversial. Some actors tried to subsume engineering under (applied) science, whereas others wanted to establish the autonomy of engineering science. The chapter concentrates on two proponents of hybrid mathematization: Robert Thurston, a leading engineer in the United States who promoted the experimental laboratories at Cornell, and, on the German side, Carl Bach, founder of the materials testing laboratory at Stuttgart and a prominent actor in the "Anti-Math Movement" of German engineering that erupted in 1895. Thus, we examine the controversial role of mathematics in prediction as a means to study the evolving autonomy of engineering.

Chapter 4 assumes a special position as a hub. The chapter covers all four modes of prediction spanning precomputer and computer times. Readers who enter the book with this chapter can move backward to precomputer times (chapters 2 and 3) or forward to the study of computer-based predictions (chapters 5 to 8). Chapter 4 exemplifies our claims with an extended case study on quantum chemistry. From its inception in 1927, quantum chemistry experienced competition between two camps that worked in a rational mode of prediction, termed *ab initio*, versus a rational–empirical one called semiempirical. Whereas the latter mode was the leading one in the 1930s and 1940s, the electronic digital computer changed the game. Quantum chemists had always utilized math tools for iterative and numerical procedures, but with the technology of the digital computer, the entire research field developed into an iterative–numerical culture of prediction that was hegemonic up to the 1980s. A strong indicator, we argue, is that the conception of *ab initio* was revised—from the rational "derived from first principles" to the iterative–numerical "computed without human intervention." However, quantum chemistry experienced another turn in the 1990s, a turn that made the notion of computational chemistry popular. We analyze this turn with a case study on density functional theory, arguably the most popular theory in recent computational science. We find that the new technology of easily available and networked PCs is coevolving not only with a new mode of prediction that we call exploratory–iterative but also with the social organization of modeling. Together they form an exploratory–iterative culture of prediction that leads to a further orientation toward predictive success at the cost of explanatory potential.

The following chapters discuss the dynamic relationship between computer technology, mathematization, and social organization in the iterative–numerical (chapters 5 and 6) and the exploratory–iterative cultures of prediction (chapters 7 and 8). Chapter 5 follows Jay Forrester (1918–2016), an engineer at MIT and a pioneer of system dynamics, who directed the development of Whirlwind, the fastest and most expensive early-generation digital computer that was integrated into the Semi-Automatic Ground Environment (SAGE) air defense system. Both the person and the machine exemplify systems thinking in which prediction assumed a pivotal role, and the numerical mode of prediction matured with mainframe centralized computing machines. Over more than a decade, Forrester expanded system dynamics from the level of an individual company to that of urban areas and to the global level. We examine the Club of Rome's "Limits to Growth" study that appeared in 1972 and built on Forrester's system dynamics. It presented a prototype for world modeling and put prediction in a policy context. Moreover, it marked a watershed beyond which computer model–based predictions gained an edge over older ways of thinking about the future. We argue that thinking about the future and political decision-making moved under the umbrella of a culture of prediction.

Chapter 6 pioneers the history and sociology of computational fluid dynamics (CFD) software. It offers a case study to show the ways in which CFD models were developed and how they moved to tackle new problems. Over the course of their travels, the models changed at many different levels from the entities they could represent to the kinds of code (both algorithms and programming languages) used to modify and add on to the models. Computational fluid dynamics models and their construction and adaptation mean different things to different users. This chapter examines this variety of social and epistemic phenomena that accompany the journey taken by CFD models.

Chapters 7 and 8 concentrate on the exploratory–iterative mode of prediction while investigating very different scientific disciplines. Chapter 7 discusses the Bayesian approach in statistics, which was a controversial topic in the philosophy of statistics but was rarely applied in scientific work. In the 1990s, however, Bayesian statistics quickly turned into a widely used predictive method in many sciences. This chapter claims that the recent upswing of Bayesian approaches hinges critically on the technology of highly available and networked computers. On their basis, we argue, Bayesian statistics joined the exploratory–iterative culture of prediction—a showcase of how mathematical and computational instrumentation interrelates with standards of rationality.

Chapter 8 is coauthored with engineer Hans Hasse, head of the Laboratory of Engineering Thermodynamics at the University of Kaiserslautern. We claim that the iterative–exploratory culture delivers predictions in ways that create new affinities between science and engineering. The present chapter exemplifies this claim with a case study on thermodynamics. Linked to the iterative–exploratory culture of prediction, important parts of thermodynamics research migrated from physics to engineering. The core of this ongoing development, we argue, is how simulations combine theory with adjustable parameters to produce predictions. This strategy of combination became possible only in an iterative–exploratory mode of modeling.[17] We find that bridging simplicity (of scientific laws) and complexity (of context of application) with the help of adjustable parameters gears science toward predictive power—and away from an explanatory capacity.

Chapter 9 offers an outlook on the future. We venture into current concepts and practices of machine learning—in particular, *deep learning* on artificial neural networks—and explore the question of whether these uses herald an emergent fifth culture of *pure* prediction. According to a common viewpoint, a good prediction redeems a claim that itself is based on other properties than merely the fact of the prediction. In some way, whatever is able to deliver good predictions has got something right about the world, or about that fraction of the world under investigation. And this something is the fundament and the true source of the predictive capability.[18] Remarkably, "pure prediction" seems to turn this upside down: prediction happens on the basis of a method, or a generic model, whose representational properties are very weak or even inaccessible. In other words, the outcome seems to be nothing but predictions.

Copyright Notice

I The Entangled History of Rational and Empirical Modes

2 Hitting the Target with Mathematics

When modern science developed in the sixteenth and seventeenth centuries, mathematics played a distinct role. Although existing accounts assign very different roles to mathematics, they treat math mostly as a homogeneous block.[1] We take a different perspective and want to pay close attention to the rich and dynamic history of how mathematics works as a tool for prediction. This history shows that there are different modes of prediction that partly oppose and challenge each other, thus turning mathematization into an endeavor marked by complementarity and tension.

We identify the rational and the empirical mode of prediction. In a first take, the rational mode assumes that mathematical laws of nature capture the world's structure and determine predictions through mathematical analysis and derivation. The empirical mode synthesizes observations or experimental data into mathematical expressions that then allow predictive extrapolation. Actually, one rarely finds historical examples that exemplify one mode in a pure form. The two modes rather resemble ideal types that stimulate and guide mathematization.[2] How and in what proportions the modes are balanced in concrete practices is a matter that scientists and engineers handle with flexibility.[3] Our examination shows that the question of when practitioners take a prediction as persuasive and when it counts as being successful is a remarkably open one—a question that different cultures of prediction address in different ways.

One of the textbook cases regarding the application of mathematics to a practical problem is ballistics. For more than four centuries, ballistics theorists have perceived mathematics as the tool that will enable projectiles to hit their targets.[4] Ballistics stands as a case for the continuous improvement of mathematics in application, one that shows that dogged determinism in applying simultaneously developing mathematical tools to evolving technological capacities yields results. And yet, with each pronouncement of greater accuracy, there has been a caveat—some claim that complete predictive accuracy is still elusive.

It is a general observation throughout this book that new technologies and mathematical instruments are used to refine predictions and increase their accuracy; but, at the same time, they open up new ways for these increasingly specialized systems to fail. Here, we exemplify this observation with the help of three episodes that unfold the history of prediction.

Episode 1 is about Niccolò Tartaglia's "New Science" of ballistics that he proposed in the sixteenth century—an early instance in which the rational and the empirical mode become discernible, although they were not fully, and definitely not clearly, separated. In episode 2, located in the seventeenth century, the two modes receive a clearer expression in the brief exchange of letters between Galileo's (master) student Evangelista Torricelli and Giovanni Batista Renieri in which they dispute the predictive accuracy of ballistics. The tension between the rational and the empirical mode culminates in episode 3 in the eighteenth century in which Benjamin Robins and Leonhard Euler both produced celebrated breakthroughs in their treatises on "new" ballistics. Robins mathematized the empirical mode of prediction.[5] Euler championed the rational mode—the essential ingredient of rational mechanics. Ballistics continued to make significant advances. We close with a brief outlook at further advances in scientific treatment, new measurement instrumentation, and new institutional organization. In particular, digital computers entered the scene and became integral parts of two new cultures of prediction. However, we do not examine the computer-related part of the ballistics story. Computer-related cultures of prediction will have center stage from chapter 4 onward.

2.1 Episode 1: With Reason, Not at a Hazard

The application of mathematics to ballistics precedes the use of gunpowder, but in the most complicated, nonexplosive cases such as that of the trebuchet, the more mathematically challenging aspects focus on the internal ballistics: the process of generating velocity to expel the projectile. However, it is undeniable that gunpowder changed the rules and aims of warfare and of ballistic science. According to Mark Denny (2011), "For 300 years after gunpowder weapons appeared, it was thought that the trajectory of the projectile consisted of straight lines connected by circular arcs" (79). This account was challenged in the first half of the sixteenth century by Niccolò Tartaglia in his 1537 treatise *La Nova Scientia*.[6]

Tartaglia (1499–1557) is a natural starting point. His treatise is the first work to frame a problem of practical relevance as a task for prediction based on scientific knowledge and mathematical tools. The case of Tartaglia illustrates how mathematization is not about the unfolding of eternal truths but about mixing and balancing heterogeneous sources and reasons, and, furthermore, about positioning the new science (including its proponents) in a social context. Tartaglia draws from a number of different traditions and produces a type of mixed reasoning that he wants to establish as a "new science." To achieve this, Tartaglia permits a baffling inconsistency: he first questions the straight lines, then uses *mathematics* to analyze the trajectory of a projectile *and* supposes straight lines to carry out this analysis.[7] The inconsistency does not so much indicate flawed reasoning, but rather is motivated by systematic reasons. Tartaglia's treatise exhibits severe shortcomings in both empirical accuracy and mathematical rigor. At the same time, his work outlines a daring construction of "mathematized ballistics" as a new object. Tartaglia involves different types of reasoning (*ragioni*) when framing this object of investigation. Without being explicit about it, he highlights the intricate tensions that hold between the different modes of math-based prediction that we call the *rational and empirical modes of prediction*. In Tartaglia's new science of ballistics, we witness the speciation of two different modes of prediction whose tension would accompany the development of ballistics and the process of mathematization for centuries to come.

From the outset, Tartaglia locates the new science of ballistics in a social context. He dedicates his work to the Duke of Urbino, who was overseeing the defense of Venice against the Ottoman Empire.[8] The political circumstances should make the new science a valuable gift.

> I fell to thinking it a blameworthy thing . . . to study and improve such a damnable exercise, destroyer of the human species . . . But now, seeing that the wolf [the Ottoman sultan Suleiman I, JL] is anxious to ravage our flock . . . it no longer appears permissible to me at present to keep these things hidden. (as cited in Drake and Drabkin [1969, 68–69])

At the same time, Tartaglia takes care to maintain contact with practitioners. His treatise answers, so the dedication reports, a question posed to him by a bombardier at the Castelvecchio six years earlier when he was working as a *maestro d'abaco* in Verona: At what inclination to the horizon would a cannon achieve maximum range? When Tartaglia chooses to put a bold claim for usefulness into his title: "The Newly Discovered Invention of Nicolò Tartalea of Brescia, Most Useful for Every Theoretical Mathematician, Bombardier, and Others, Entitled *New Science*," he is appealing simultaneously to patron, theoretician, and practitioner.

The new science was about amplifying established opinions and goals, not about disruption. Many historians take this as inherent conservatism. We would like to point out that Tartaglia is framing the question in a new way. Above all, the new science is going to be about prediction. Tartaglia does not merely answer the bombardier but projects the bombardier's task as being a part of a comprehensive science-based practice. This is palpable when looking at the tool that Tartaglia invents for this task, the squadra (see figures 2.1 and 2.2).

The squadra was designed as the instrument to measure the angle of elevation of a cannon barrel so that it could be adjusted. This angle, in turn, is of value only when the bombardier has a mathematical procedure at hand

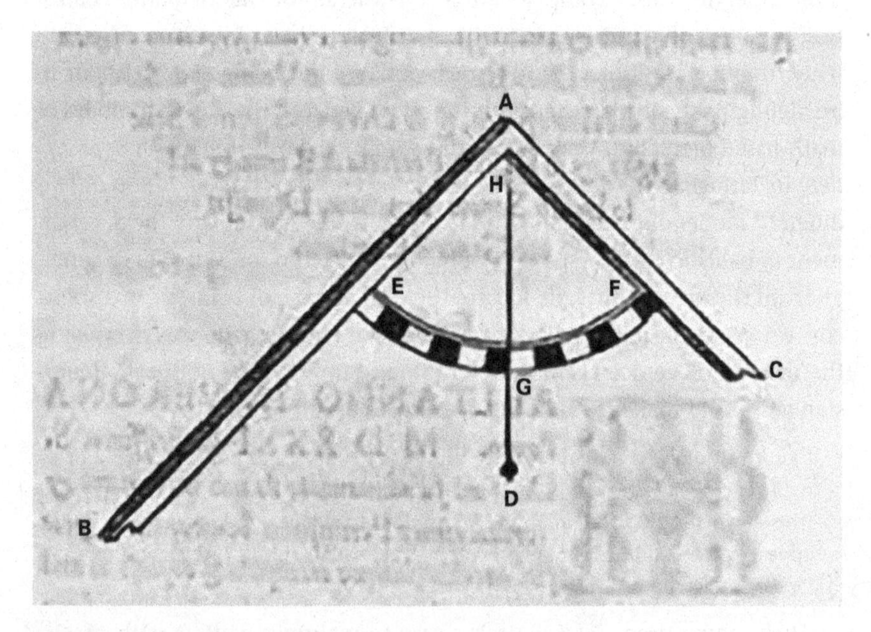

Figure 2.1
The squadra as depicted in Tartaglia's *Nova Scientia*.

A peece mounted at 6. points or 72. minutes.

Figure 2.2
A cannon with the squadra at 45-degree elevation as depicted in Tartaglia's *Nova Scientia*.

that relates the (measured) angle to the target. The bombardier's question about maximum range turns out to be a special case of this problem. One characteristic of a culture of prediction is that concrete instances of prediction are embedded into a general practice of solution. Tartaglia acknowledges the value of experience but insists that ballistics as a science (his invention) proceeds from experience to general rules.[9] And mathematics is the central element in scientific reasoning about prediction. When gunners act in line with the new science, so Tartaglia promises, they will aim at their targets "with reason, not at a hazard."[10]

But what is reasoning about a trajectory supposed to look like? At the time, this question could not be answered in a straightforward way; furthermore, it was unclear what it refers to. Thus, Tartaglia's prime achievement lies exactly in constituting the trajectory as an object of inquiry. By mathematizing the trajectory of a projectile, Tartaglia turns it into an entity open for investigation. Consequently, the nature of this investigation depends on the available mathematical tools. Tartaglia invokes four different kinds of reasonings (*ragioni*): physical and geometrical, demonstrative geometrical, Archimedean, and algebraic. The wide spectrum of reasonings is remarkable because it transcends disciplinary boundaries. Arguably, his background as *maestro d'abaco* was conducive to this versatility. Teaching practical applications of algebra, arithmetic, and geometry to artisans made him inclined to amalgamate mathematical approaches usually kept separate in the higher educational system.[11] "In contrast to university and humanist mathematics, which focused on abstract, rigorous demonstrations, the aim of the abacists was to teach concrete problem solving" (Ekholm 2010, 185). For instance, Tartaglia describes how he was employed to calculate tables that bakers could use to determine the cost of bread on the basis of the cost of ingredients (Ekholm 2010, 201). Much like a baker would want to have a formula that connects the type and amount of dough to the fermentation and baking time, the bombardier wanted a formula connecting the angle of elevation to the distance of target to be hit.

Tartaglia uses algebra[12] in his treatise on ballistics as if it were common practice for abacists: naming unknowns and then determining their value by means of equations. Tartaglia fuses this with geometry when he multiplies (unknown) sides to calculate areas—something absent from Euclidean geometry. Another type of reasoning mentioned by Tartaglia is geometrical–demonstrative. The *Nova Scientia* clearly mimics Euclid's *Elements*—at that time, the gold standard of books on mathematics—when it employs its theorems to proceed by geometrical demonstrations.[13] Tartaglia is not fully successful in imitating Euclid—his demonstrations lack mathematical rigor and conclusiveness. At crucial points, Tartaglia alludes to experience but without relying on systematic experimentation. The aim of producing numerical values for practical problems by whatever mix of mathematical techniques stood in stark contrast to what was taught in higher education. Furthermore, he still drew on Aristotelean and impetus theories of motion.[14] Whereas Tartaglia is usually criticized for this methodological mess, we would like to reverse the perspective.

The mathematics of the *Nova Scientia* is indeed a "strange amalgam" (Ekholm) of abacus, axiomatic–deductive, and natural–philosophical traditions. Tartaglia treats none of these ingredients flawlessly. But this sort of critique fails to appreciate Tartaglia's main achievement. Through this amalgam, he can propose an agenda toward prediction that links the artisan knowledge of the bombardier to both scientific knowledge and the duke's interest. It is important to bear in mind that mathematical tools themselves had to evolve to become part of a practice of prediction.[15] The problem and the tools coevolve when Tartaglia constructs the trajectory of a projectile as a new object of scientific investigation. To be clear, this culture of prediction—what bombardiers need to know and to do in order to shoot with reason and not at a hazard—was more vision than reality.

The frontispiece of the *Nova Scientia* depicts Tartaglia's vision in a rich and telling way, with education, practice, mathematical disciplines, the sciences, Aristotle, Plato, and Tartaglia himself (see figure 2.3) present in the picture.

> We are shown an enclosure encircled by a high wall. Steps are depicted mounting up to a single gate guarded by a venerable man entitled *Euclide*. Only by this discipline is one admitted to the sciences within the enclosure. (Strong 1936, 57–59)

On the left of Euclid, who is acting as doorkeeper, one individual is trying to enter the enclosure by his own means, but his ladder is too short.

Figure 2.3
The frontispiece of Tartaglia's *Nova Scientia*.

There is no way to shortcut mathematical education. A second opening of the enclosure is on the opposite side of the Euclid gate. Some steps lead up to *Philosofia* depicted as reigning on her throne in solitude. Actually, the opening is more toward rather than leading away from the central enclosure. Philosophers have to move down to become involved in science. Both Aristotle and Plato stand on the steps, the former a bit closer to the arena of science. The sciences form a chorus of women, each carrying a label, among them *Geometria, Astronomia, Arithmetica, Musica, Hydromantia, Geomantia,* and *Architectura*. In the front row amid this chorus stands Tartaglia himself. He has no label, but the entire enclosure, spread out as a secluded place between philosophy and normal life, appears to be the arena for his new science. The other sciences marvel at it. The enclosure is a kind of shooting ground with a cannon and a mortar firing projectiles whose trajectory is depicted very plainly.

The frontispiece locates the new science among existing sciences, but, at the same time, it opens up new territory. Plato holds the banner *NEMO HUC GEOMETRIE EXPERS INGREDIATUR* [Only knowers of geometry are allowed to enter], and thus mirrors the figure of Euclid. In between them, there is a space for mathematized science that is concerned with practical matters. This is the new invention and discovery Tartaglia is proud of, or better, that the frontispiece illustrates. Even the new object of investigation—that is, the trajectory of a cannon ball—is included in the illustration. Remarkably, the trajectory on the frontispiece is continuously curved, and Tartaglia is counted as the first to question the common "straight line plus arc" wisdom. However, unlike what the frontispiece displays, this treatise actually models the trajectory as straight lines plus arc. For what reason would he allow such blatant inconsistency?

We argue that the reason lies in the claim for a predictive practice. In such a practice, mathematics has to be conceived as a tool. For a tool to work, a number of conditions have to be met, including acceptance, tractability, and usability. Tartaglia was probably well acquainted with the parabola because, after all, he did publish the translation of Archimedes's quadrature of the parabola. However, much like the frontispiece suggests, the only way to ballistics is through Euclid. At the time, demonstrative reasoning and analysis demanded working with the geometrical elements known from Euclid. Consequently, Tartaglia moves from the physically adequate curved trajectory to a mathematically admissible straight one. He acknowledges that his geometrical model is only approximately valid.

> No trajectory other than the perpendicular can have any part that is perfectly straight, because the weight of the body continually acts on it and draws it toward the centre of the world. Nevertheless, we shall suppose that part which is insensibly curved to be straight, and that which is evidently curved *we shall suppose* to be part of the circumference of a circle, *as they do not sensibly differ*. (Tartaglia [1537, Book II, Supposition 2] as cited in Ekholm [2010, 190], emphasis added)

According to Tartaglia's account, mathematization does not automatically result in a representatively adequate model but in a model whose flexibility can be used to adjust it in an (predictively) adequate way.[16] However, predictive adequacy was a mere postulate, neither checked nor correct. In fact, Tartaglia argues about the form of the trajectory—straight upward according to angle of shoot, circular arc of certain dimensions, finally perpendicular descent toward earth—so that he can break down the problem into established building bricks and infer that an elevation of 45 degrees will maximize the range.

Hence, for Tartaglia, the new science depends on a delicately balanced mixture. The agenda of prediction must be promising enough to interest the duke, it must be in line with the experience of practitioners (the elevation of 45 degrees, halfway between vertical and horizontal, was widely held to result in maximum range), and the mathematical instrumentation must be academically accepted (Euclidean geometrical shapes and forms of analysis). One can denounce the willingness to respect all these conditions as opportunistic. Moreover, there can be no doubt that predictions along the lines of Tartaglia's treatise are of staggeringly low empirical adequacy. The success and fame of Tartaglia's ballistics do not rely on the accuracy of the predictions. At the time, this accuracy could not have been assessed anyway. Rather, what makes Tartaglia influential is that he was the first to create an amalgam that constructed ballistics as a predictive science.

In Tartaglia's amalgam, we can discern two modes of prediction, even if they are not elaborated in any clear way. The first can be called the rational mode. It adopts mathematics to grant generality and authority in deriving predictions from general laws. The second mode we call empirical. It dwells on the flexibility of mathematical tools that can be adapted to practical situations. The next episode shows how the work of Galileo and his student

Torricelli advanced mathematization and further brought to light the two modes of prediction.

2.2 Episode 2: Early Expression in the Seventeenth Century

Tartaglia's work was important to Galileo (1564–1642), born seven years after Tartaglia's death. Galileo has a protean reputation for being the inventor of the experiment in science, for being a theoretician who advocated for the superiority of mathematization, and for being an engineer who actualized theory in technology.[17] Notwithstanding the many reasons that distinguish Galileo as a pioneer, there are continuities with Tartaglia. Both belong to the Italian tradition of mathematized mechanics.[18] On the social side, both men undertook constant efforts to gain patronage.[19] Importantly both also took up the "Archimedean" approach, which combined physical and mathematical reasoning.[20]

Compared to Tartaglia, however, Galilei introduces a much more elaborate vision of what new sciences based on prediction would look like. Like Tartaglia, he hails ballistics as a "new science," though for Galileo, this means that he achieves what Tartaglia had merely dimly intended. We start with a brief account of Galileo and then turn to the dispute between Torricelli, who elaborated and defended Galileo's account, and Renieri, who complained about its lack of empirical adequacy. This dispute brings to the fore how wielding mathematics as a tool for prediction involves both the empirical and the rational mode of prediction.

In his 1638 tract, *Discourses and Mathematical Demonstrations Relating to Two New Sciences*, Galileo introduces ballistics and the strength of materials as new sciences.[21] The fourth day of the dialogue between Salviati (the Copernican academician) and Sagredo (the educated layman) is devoted to ballistics and promises to exploit scientific knowledge to make predictions. Galileo includes several tables describing projectile trajectories for a variety of different initial angles (i.e., elevations of a cannon muzzle). A point of fundamental importance that distinguishes him from Tartaglia is that Galileo starts from a new theory of motion and derives the exact mathematical form of the trajectory from it. The projectile, so Galileo's pathbreaking theory, is moving in the direction taken by the cannon barrel, whereas, at the same time, it is constantly affected by a force toward the center of the earth. Tartaglia had no mathematical concept of such a theory and consequently could not derive the trajectory from theory in any mathematical way. Famously, Galileo describes the trajectory of a projectile as a parabola, and the tables in *Two New Sciences* show the altitude and sublimity of each initial angle. However, Galileo's geometrical methods do not give him a general formula for the relationship between angle and range.

Curious findings appear in the table—and both Mersenne and Torricelli subsequently took up the table and tried to refine it not only to produce more useful information for gunners but also to make it mathematically more rigorous. Their real interests, though, lay in the latter, despite lip service to utility. In fact, the goals of rigor and utility, conditioned on the available mathematics, were at odds with each other.

The notion of mathematical rigor changed considerably over the course of history. Galileo used mainly geometrical methods, and it was the *more geometrico* standard of Euclid that defined rigor in demonstration.[22] The point for Galileo is that the theory of motion should be sufficiently mathematized so that essential aspects of physical processes—such as the trajectory of a projectile—can be derived by mathematical reasoning. We deem this point of crucial importance: it shows that mathematization must aim at a *balance*. On the one side, a theory must be simple enough that one can carry through the derivation with the mathematical means at hand. On the other side, it must be sufficiently rich in content so that one can derive relevant quantitative consequences from it. Achieving a balance might not be possible at all; or, if possible, there might be different and conflicting balances. Distinguishing modes of prediction is a helpful tool when it comes to analyzing the balances attempted by different actors.

What kind of balance does Galileo strike? He does not fully stand up to the promise of the rational mode because he cannot offer a general solution to the question of the trajectory. Although he is able to derive his famous result that the trajectory is a parabola, and also prove that the 45-degree angle therefore produces the maximum range, he is unable to present a general account that would tell the gunner how to predict the range.[23] And Galileo does not provide a prediction in the empirical mode either because his tables are geometrical demonstrations rather than real predictions for gunners. Most significantly, because Galileo relies on standard geometrical means, he can tackle the ballistics problem only if air resistance is left out.[24]

The crucial question then is whether ignoring air resistance matters in terms of the actual prediction. If not, the rational and empirical mode would yield congruent results; rigor and utility would be in accordance. Galileo is well aware that he has no viable alternative to ignoring air resistance, but he also suggests that this would not matter in terms of the prediction. This suggestion shows his faith in the rational mode. Here is how he deals with it.

In Proposition 7 of the *Two New Sciences*, Galileo establishes that the 45-degree angle results in maximum range.[25] Sagredo and Salviati then discuss the relative merits of mathematical against experimental validation. Sagredo, the educated layman, favors the former: "The force of necessary demonstrations is full of marvel and delight; and such are mathematical [demonstrations] alone." Salviato, the Copernican academician, agrees: "The knowledge of one single effect acquired through its causes opens the mind to the understanding and certainty of other effects without need of recourse to experiments" (Galileo 1974, 245).

Although Galileo praises the force and generality of mathematical demonstrations, he assigns an important role to experimental knowledge. Namely, "the knowledge of one single effect through its causes" is the starting point on which the generalizing mathematical representation is based. In Proposition 1 of the Fourth Day, Galileo reports empirical knowledge that a projectile fired horizontally in the air follows a path that is a semiparabola. From there, Galileo argues, one can move over to the mathematized theory and derive further knowledge. He acknowledges, however, that such a move might be provisional because addressing a new situation might call for additional experimental coverage. Galileo here navigates between the commitment to experiment and the promise of the rational mode of prediction.

Galileo acknowledges the importance of the question of whether the presence of air resistance makes a significant difference. The parabola has led him to the maximum range Proposition 7. The parabola form required the no-air-resistance condition. Without it, the entire endeavor would be nil. At this point, Galileo refers to the common knowledge among practitioners. They knew empirically, Sagredo said, that 45 degrees is the correct result:

> I already knew, by trusting to the accounts of many bombardiers, that the maximum of all ranges of shots, for artillery pieces or mortars—that is, that shot which takes the ball farthest—is the one made at elevation of half a right angle . . . But to understand the reason for this phenomenon infinitely surpasses the simple idea obtained from the statements of others, or even from experience many times repeated. (Galileo 1974, 245)[26]

Was this (alleged) common knowledge of practitioners good enough to work as a basis for neglecting air resistance altogether?

> In projectiles that we find practicable, which are those of heavy material and spherical shape, and even in [others] of less heavy material, and cylindrical shape, as are arrows, launched [respectively] by slings or bows, the deviations from exact parabolic paths will be quite insensible. (Galileo 1974, 276)

Galileo knows that, with the geometrical means available to him, he can derive the parabola only when neglecting air resistance. Without ignoring air resistance, his new science of ballistics would have been a nonstarter. In other words, Galileo's claim for prediction rested on his faith in the rational mode. Such a claim is plausible (only) in a rational culture of prediction in which this faith is shared.

At the time, the rational culture was by no means firmly established. Torricelli, with whom this episode continues, had to struggle with how to defend Galileo's prediction. Evangelista Torricelli was probably Galileo's most ardent pupil. "Among Galileo's leading followers only Torricelli seems to have been in general agreement with the master on these points and may therefore be taken as his best representative" (Segre 1991, 91).[27] He elaborated Galileo's theory of ballistics, turning it into the general predictive theory that Galileo arguably had strived for. Torricelli published his findings in 1644 in his *Opera Geometrica*, in the second section "*de Motu*."

Torricelli believed that gunners were dismissing Galileo's table because it not only offered little advice on calculating accurate trajectories but, according to Segre, it also contradicted their conventional understanding that up to 45 degrees, projectile ranges would increase in direct proportion to the angle of elevation. Torricelli sought to make an improved table that followed Galileo's but allowed an empirical correction for air resistance at high velocity. High velocities were achieved by cannons; mortar shots, like nonexplosive weapons, were considered low velocity by the two men.[28] The claim was that the range was proportional to the sine of 2α, where α was the angle of elevation.

Thus, Torricelli was able to express in a formula what the parabola meant in terms of prediction—and he alluded to Tartaglia's predictive agenda by designing a squadra that made the predictive mathematics directly usable (see figure 2.4)—a clear statement that ballistics, first, was now on a sound scientific footing and, second, had become the useful predictive instrument that gunners had sought for so long. Or so it would seem.

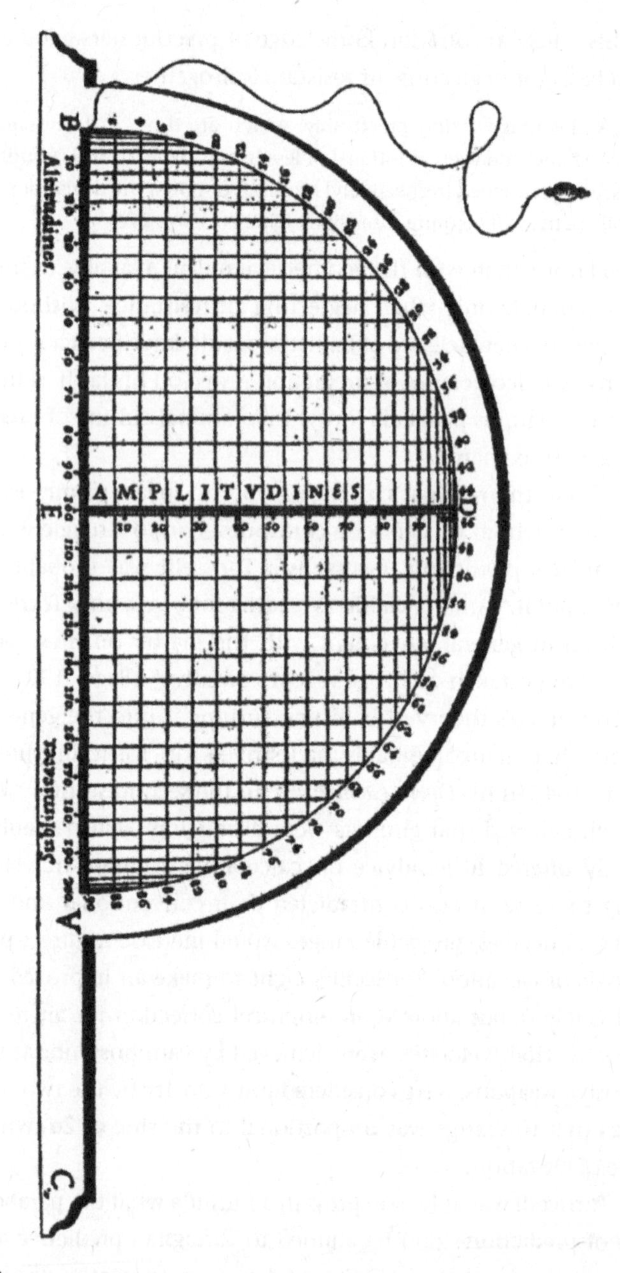

Figure 2.4
The squadra from Torricelli (1644, 235) that correlates (predicted) range and elevation.

Torricelli defended Galileo's position on various occasions. One part of the criticism targeted the theory of motion and, hence, was linked only indirectly to ballistics. People such as Mersenne, Descartes, or Roberval did not accept that gravitational force would constantly accelerate a body toward the center of the earth. They put forth empirical observations indicating that falling bodies attained maximum speed. Torricelli reacted by stressing the mathematical coherence of the theory—up to the point when he stated that the matter of true or false would not be important.

> I do not care whether the principles of the *De motu* are true or false. For if they are not true, let us feign they are true, as we have assumed, and then look at other speculations derived from these principles, not as mixed but as purely geometrical. (Torricelli in a letter to Ricci, as cited in Segre [1991, 93])[29]

The conflict did not spare ballistics. In 1647, Giovanni Batista Renieri in Genoa wrote to Torricelli expressing concerns about Galileo's and Torricelli's results. Based on empirical tests,[30] Renieri doubted Galileo's parabolic account. "I was astonished that such a well-grounded theory turned out to work so badly in practice" (English according to Segre [1991, 94]). Renieri, as opposed to Torricelli, was far more experienced in working with both cannons and the gunners who use them. For Renieri, ballistics was a real problem and not a philosophical puzzle.

Torricelli probably cringed but kept steadfast to his standpoint that the rational and empirical modes were in harmony. The appeal of ballistics as a science depended on it being simultaneously both useful for gunners and a challenge for natural philosophers. Torricelli wanted to preserve the core scientific result—that is, that the mathematical form of the trajectory is a parabola. Starting from the parabola, Torricelli had been able to provide the sine 2α formula relating angle and range, cashing in exactly what the predictive agenda had promised. Renieri pointed out, however, that the predictions were apparently off the mark. Scientific "grounding" and predictive accuracy were less in harmony than the Galileo–Torricelli account of ballistics wanted to have them.

Torricelli still did not fully abandon the supposed agreement between the rational and the empirical mode.[31] He hoped that refined experimental tests would vindicate the theory. When Torricelli wrote back to Renieri with specific instructions for carrying out experiments, Renieri diligently followed them but still did not come close to the ranges predicted by the tables. Another of Torricelli's defensive responses to Renieri was that in *"de Motu,"* he was addressing philosophers, not gunners (which is obvious because it would be unlikely for gunners to have either the skill or the time to read Torricelli's Latin treatise).

For Torricelli, the retreat into the purely geometrical is acceptable as a philosophical move; for Renieri, this is pointless. Torricelli resorts to the rational standpoint: the mathematical derivation from theory is more important and opens up the potential for predictions; their accuracy is a secondary problem. Renieri, in contrast, sticks to the empirical mode of prediction. Advancing mathematical treatment is valuable but under the condition that it increases predictive accuracy.

For Segre, this episode is exemplary of the tension and conflicts between early modern philosophical and artisanal traditions, and the challenge of having any sort of dialogue between them. He argues that while Torricelli created the impression that he was working experimentally on these matters, in fact his assumptions and suggestions in the letters to Renieri about the possible precision of cannon technology revealed him to be a novice in working with cannons. Indeed, A. R. Hall, in his *Ballistics in the Seventeenth Century* (2009), also comments on the immaturity of engineering and technology in seventeenth-century weapons, claiming that they were simply not standardized or reliable enough to support a science of gunnery.[32] Consequently, the empirical mode of prediction at that time was de facto an illusion. Neither did the relevant empirical data exist nor would it have been possible to obtain them—despite the hopes and suggestions of Tartaglia, Galileo, and Torricelli.

Here the episode raises the question of the tool character of mathematics. Must the tool produce something practical? If so, then accuracy of prediction is the primary evaluation criterion. If the aim is to accurately predict trajectories and ranges, failing to do so must render the tool useless. But if the goal is to describe the geometry of a trajectory *in theory*, then Torricelli's use of mathematics is much more satisfactory. This is an instance of how the problem of prediction and mathematical tools codevelop.

On another note, the correspondence between Torricelli and Renieri also introduces the idea of "mixed mathematics," a term that captures what we have found in Tartaglia's work in a much more unsystematic fashion. In the seventeenth century, there was no term for applied math; applied math problems such as ballistics were referred to as "mixed mathematics." All sorts of engineering problems were described by this phrase, yet what was being "mixed" is open to some speculation.[33] We have argued that when pursuing the predictive agenda of ballistics, researchers were seeking a balance between the two mathematics-based modes of prediction. We have also argued that with the type of data and the mathematical tools available at the time, no balance or mixing could have fulfilled the hopes articulated by ballisticians. The next episode in ballistics jumps forward to the "classical" eighteenth century when the two modes of prediction parted ways and two cultures—a rational and an empirical one—matured.

2.3 Episode 3: Robins, Euler, and Conflicting Modes of Prediction

The third episode takes place about one century later in the 1740s. A lot happened in that century that is relevant for the development of mathematics and ballistics. We focus on formulating an argument about the two mathematics-based modes of prediction that then reached full bloom.[34] We see a significant further co-development of problems and tools that puts the empirical and the rational mode alongside each other. The leading characters in this episode are Benjamin Robins (1707–1751) and Leonhard Euler (1707–1783).

Robins and Euler both had multidimensional and peripatetic careers, with the English Robins dying in India and the Swiss Euler in Russia. Each contributed to many different projects, many of which were extensions of Newtonian mechanics. Both were particularly interested in producing a general solution (or a prediction) of the kinematics of high-speed cannon shots. Whereas Robins was the more practically oriented of the two, he was no mathematical slouch, either.[35] At the age of twenty, he published a proof of Newton's quadrature in the *Philosophical Transactions* that got him elected to the Royal Society the same year. For most of his career, Robins worked as what we would today term a "civil engineer"—designing bridges, draining fens, planning harbors, and so forth. In his 1742 treatise, *New Principles of Gunnery*, Robins (1972) undertook to turn ballistics into a mathematized science with predictive power. He set up a comprehensive agenda that included both empirical and theoretical research and that can be seen as the first full account of the empirical mode of prediction in ballistics. Robins expressed this in the subtitle of the *New Principles* "containing the determination of the force of gunpowder and an investigation of the difference in the resisting power of the air to swift and slow motions."

Robins synthesizes research from very different directions. He extends specific propositions from Newton's *Principia* to ballistics. The *New Principles* offers a sophisticated account of the internal ballistics of cannons, a description of the production of gunpowder, and the measurement of various ranges and trajectories that he summarized in a set of tables. Robins also designs an instrument, the ballistic pendulum, for measuring the muzzle velocity of a projectile. He is aware that predicting an (external) trajectory requires accurate knowledge about the initial conditions, especially muzzle velocity. And this input can be controlled only when gunpowder is standardized—hence his work on gunpowder. Robins sees another lacuna, namely measuring air resistance that is a—possibly major—factor determining the trajectory. To fill this lacuna, he invents another instrument called the "whirling machine." These two instrumental developments provide new data. Taking account of the empirical findings led to new methods of calculating projectile trajectories involving new equations and algorithms. In his *New Principles*, he writes:

> This has principally given rise to the ensuing treatise, in which the force and varied action of Powder is so far determined, that the velocities of all kinds of bullets impelled by its explosion may be thence computed, and the enormous resistance of the air to swift motions (much beyond what any former theories have assigned) is likewise ascertained. (4)

Unlike Galileo's or Torricelli's tables, Robins's tables are designed specifically for use by gunners. One of the interesting problems Robins acknowledges in his work is that his experiments often generated more variation between different trials than between trial and theory. In 1743, Robins had no statistical theory on which to rely when managing his data and determining which trial results should be either relied on or discarded. But his observation of the problem stands out in contrast to the work of ballistics theorists, none of whom mention it as a challenge, and with some such as Torricelli seemingly very naïve about the inconsistency of real weapons in the real world. Clearly, the empirical data play a guiding role in Robins's work. He calculates a ballistics table with mathematical means, but producing data and inventing measurement technology makes up a large part of his agenda. Mathematics is needed to adapt data, technology, and theory to each other.

Robins's work shows in a paradigmatic way how problems and tools coevolve in the empirical mode of prediction. One part of his research concerned the effects of rifling. Robins used wooden bullets in rifled barrels to show that the projectile had greater accuracy. He used a lead ball to show that rifling increased the projectile's range. He discovered that the way the ball exits the barrel causes the ball to rotate, and this disturbs the uniform air flow over the ball. Air flows faster on one side than the other, and the ball deflects toward the higher-velocity side (in the direction of the spin). Robins proposed that the rifled bullet's direction of forward motion coincides with its axis of motion. That is, Robins developed a mathematical account of what later became known as the Magnus effect.

Robins also invented measuring instruments. With the help of his "whirling arm," Robins could confirm that air resistance varies roughly as the square of the speed for slow-moving projectiles. This agreed with what Newton, Johann Bernoulli, and others had assumed. However, he observed a sharp increase in resistance (by about a factor of three) at supersonic speeds. This was unanticipated and not accounted for by Newton and others.[36] Robins was uncertain about how to treat this surprising finding because it seemed to escape the standard expectations of ongoing mathematization.

Was his work an application of Newtonian mechanics? Here, the notion of "application" is precisely what is problematic. In his treatise, Robins does not make a theoretical contribution to mechanics—nor does he further develop the mathematical tools. Robins's calculation of air resistance certainly does not violate Newtonian mechanics, but Newton did not cover how air resistance changed in complex ways relative to speed. There is a well-charted history of Newton suggesting that air resistance was proportional to the square of speed but being unable to derive the consequences for the trajectory. Johann Bernoulli, who did not use Newton's fluxions but the more algebraically framed Leibnizian calculus, was able to compute the range—and was proud that the "continental" calculus proved to be the superior mathematical instrument.[37]

But Robins approaches the matter differently and not as a competition between mathematical tools, skills, or elegance when bringing phenomena under general laws. His air resistance calculations are motivated by empirical data, and his aim is to develop a formula that would be in accordance with his empirical findings. Thus, Robins establishes the empirical mode of prediction through his predictive agenda and achievements in measurement technology.

In 1747, Robins received the Copley Medal from the Royal Society in recognition of his achievement in developing the new science of ballistics.

Published in 1742 as *New Principles of Gunnery*, his work on the predictive agenda was a huge success. It was taken up as a book at Woolwich, translated into French by LeRoy for use in their artillery schools, and, most famously, read and commented on by Leonhard Euler who published the work in German in 1745 with a voluminous additional commentary and mathematical extension.

Euler pursues a different approach that made him the outstanding leader of "rational mechanics." He received a copy of the *New Principles of Gunnery* shortly after its publication in 1742 and began to write a commentary and translate it into German (published 1745). He published a further commentary in the *Memoirs of the Royal Academy of Berlin* in 1753 (Euler 1922b). Euler's work not only extends the mathematical principles put forth by Robins but also makes some changes. We want to highlight that Euler sees mathematics in a different role and strives for a fundamentally different balance between the modes of prediction. Euler delivers major achievements in elaborating the mathematical tool of calculus. In fact, objects such as Newton's equations received what is nowadays their common form through the hands of Euler. For him, mathematics serves as the backbone of rational mechanics. Consequently, he tries to develop a tool that will enable rational mechanics to cover relevant cases. At the same time, which cases are deemed scientifically relevant is determined by whether analytical methods could cover them. This is a typical instance of the co-development of tool and problem—a dynamic that separates Euler's and Robins's accounts from each other.

Euler was a leading figure in eighteenth-century mathematics and science.[38] For Clifford Truesdell, Euler was a hero of rational mechanics.

> Mechanics is a science of experience; for the theorist, physical experience is balanced against the experience of earlier theories for the phenomena. The history of rational mechanics is neither experimental nor philosophical; it is *mathematical*; it is a history of *special problems*, concrete examples for the solution of which *new principles and methods* had to be created. But the solution of the special problem was never left to stand alone; since there was only one true mechanics, the special case served not as an end in itself but as the *guide to the right conception*. (Truesdell 1968, 96)[39]

Scholars have understood Robins's *New Principles* and Euler's translation very differently. Clifford Truesdell (1954) dismisses Robin's work as a mere application of Newtonian mechanics and refers to the *New Principles* as "a little budget of rules, experiments, and guesses," whereas he calls Euler's "great treatise on ballistics . . . the first substantial work on the subject" (xxxviii). According to this viewpoint, Robins's work became scientific only through Euler's annotations. Steele is of the opposite opinion: his hero is Robins who made ballistics a science that is relevant to practitioners.[40] We maintain that juxtaposing the two men as heroes of different agendas misses the point. Both men set up a predictive agenda that assigned a central function to mathematical tools. Both men attacked their agenda in a masterful way. However, their work was defined by different modes of prediction. And the contrasting assessments of their merit mirror the (implicit) preference for one of these modes.

Euler (1922a) praises Robins's experimental results but also acknowledges that the true trajectory (i.e., its mathematically described shape) still remains not well known (prop VI, chapter 2). He sets himself the task of determining the trajectory—a showcase for the power of his mathematical methods. But they are not almighty, so Euler has to simplify the equations that Robins had come up with. Most importantly, Euler assumes that air resistance depends on the square of the velocity—instead of square plus fourth power as Robins had proposed. This falls back on Newton and Huygens, though neither of them had been able to treat the case mathematically in such a complete way as Euler. He calculates approximate trajectories based on Bernoulli's solution. Euler formulates second-order differential equations that present some of the earliest instances of the way such equations are conceptualized today. He obtains equations for range, altitude, and velocity in both the ascending and descending part of the trajectory—the vertical descent as limiting case of the descending part, and the positive slope as asymptote for the ascending part. "As a demonstration of the ability of analytic methods to solve difficult physical problems of practical interest, the work was a masterpiece" (Barnett 2009, 100).[41]

The likes of Truesdell agree and take this appraisal as an indication of its overall scientific quality. But on the basis of Robins's data, Euler's predictions are still not a good match. Euler was surely aware that something like the fourth power of the velocity entering the equation would have rendered the entire problem intractable. He decided to go for the solvable problem—that is, to define the problem of prediction so that it could be tackled by his mathematical methods.

Here is a second example that highlights where Euler and Robins differ. Robins had observed the (Robins-)Magnus effect that influenced the trajectory. Euler discarded the effect and held imperfections in bullet curvature responsible. He allowed that these kinds of imperfections could lead to deviations between empirical observations and prediction, but they could be plausibly left out of the picture because they are concerned with questions of design and standardization. Rational mechanics is not in charge of dealing with design imperfections. Had Euler accepted the effect as part of the physical problem, he would not have been able to derive an asymptotic trajectory. Being able to derive predictions—in the rational mode—hinged on *excluding* the Magnus effect.

Like Torricelli and Renieri, Euler and Robins were using mathematics as different kinds of tools for different purposes and, most especially, for different audiences. Euler's audience was not gunners, as evidenced by the venues in which the work was published. Euler was not using mathematics as the same kind of tool as Robins, although the mathematics included in their two treatises were very similar (remembering that, in fact, this piece of Euler's work was merely a translation of Robins with commentary and "corrections"). Both could claim great success in their predictive agenda.[42] However, they framed the concept of prediction differently—in the empirical respectively rational mode. For Robins, the empirical data were the basis, and mathematization should create a practicable tool able to recover these empirical data—and, of course, produce more predictions. For the empirical mode of prediction, the flexibility of mathematics is a crucial ingredient. Euler, in contrast, was led by the adequacy of mathematics as the tool to elevate rational mechanics. Consequently, he did not invest in the empirical side but tried to develop the calculus into a machinery for solving differential equations and thus broaden the scope of rational mechanics.

2.4 Toward Cultures of Prediction

The modes of prediction became central pieces in what we call cultures of prediction. Such cultures comprise more than modes of prediction. They need established practices of prediction that are used (and learned) in a socially organized and institutionalized way. The episodes here in chapter 2 have brought us to the point where the rational and the empirical modes of prediction are clearly discernible. The development of the rational and the empirical cultures of prediction begins and gathers momentum. When Truesdell looks back and describes rational mechanics, he is in a way promoting the rational culture and also projecting it back into the eighteenth century. This book introduces and discusses cultures of prediction in all of the remaining chapters, each culture with a mode of mathematical prediction, and each one entangled with the discourse on science and engineering.

We could have continued with episodes illustrating how mathematical tools coevolved with predictive problems. However, we shall not do this for two reasons: first, the mathematical side becomes quite technical, and, second, toward the twentieth century, two important things happened that deserve treatment in their own chapters. One is the hybridization of the empirical and the rational mode. Such hybridity proved to be pivotal for predictions but was deeply contested, mainly because of the way it played into the relationship between science and engineering. Therefore, we devote chapter 3 to the debated role of mathematics in late nineteenth-century mechanical engineering. The second important reason is that computation and numerical approximation emerged as new subjects of mathematical theory and practice. This led to new cultures of prediction (we discern the iterative–numerical and the exploratory–iterative) that are discussed in chapter 4 and all later chapters.

We conclude this chapter with a brief outlook at how the rational and the empirical modes continued to shape the development of ballistics. In 1852, Gustav Magnus published "Über die Abweichung der Geschosse: Und über eine auffallende Erscheinung bei rotierenden Körpern" [On the deviation of projectiles: And on a remarkable phenomenon in rotating bodies] in the journal *Annalen der Physik und Chemie*. Thirteen years earlier, Siméon Denis Poisson published his *Recherches sur la Mouvement des Projectiles dans l'Air* [Research on the movement of bodies in the air]. In his 1893 treatise, "On the path of a rotating spherical projectile," Peter Tait's comment is that "Poisson's treatment of the subject is unnecessarily prolix, and in consequence not very easily understood." Thus, it is apparent that the mathematics of ballistics had arrived among the community of physicists in the nineteenth century, even though there remained disagreement about the treatment. Tait was a Robins apologist; he compared Poisson's treatment to Robins from about a century earlier. Tait also claimed that Magnus did not really know the work of either Newton (on this point) or Robins. Tait pointed out that Magnus also disparaged Robins's work, writing that Robins generally believed the trajectory-diverting effect was produced by the spinning projectile. But the phrasing makes it clear that Magnus considers this of little influence in his own mathematical description. Magnus proved what Robins merely believed.

Moreover, Magnus was the first to outline a mathematical prediction. Robins's was a demonstration of an empirical phenomenon that he did not translate into prediction—in part, due to the rarity of rifled weaponry that would make such a prediction useful to gunners. Magnus's work yielded the predictive results that earned the phenomena its name (see figure 2.5), but Magnus's calculations were hardly aimed toward gunners. They redefined the scientific frontier of fluid dynamics.[43]

For Magnus and Poisson, the question how the spinning of a projectile affects its trajectory remained not only an interesting practical problem for military engineers and gunners, but also constituted a respectable mathematical concern. For these mathematical physicists, accurately predicting the trajectory of a projectile was perceived as a true scientific challenge and not, as Torricelli would have had it some two hundred years earlier, a matter of little intellectual interest, dismissed with a claim that mathematics was not a tool for describing truth.

In a sense, Magnus has claimed ownership of the calculation of the trajectory of a spinning projectile by giving his name to the effect. It is equally clear that Newton described the effect in the seventeenth century, and Robins both observed it in his tests and explained it verbally. But, in this case, it

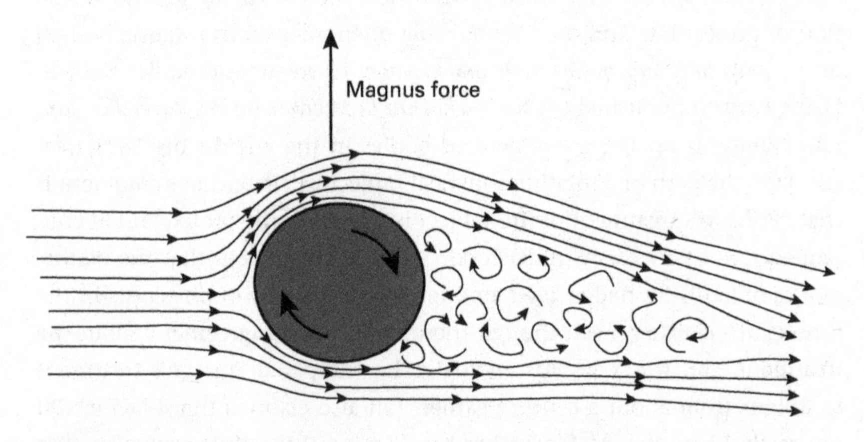

Figure 2.5
A sketch of the Magnus effect with streamlines and turbulent wake. Courtesy: Wikimedia commons.

was mathematical calculation of the trajectory incorporating the effect that conferred the naming rights.

Ballistics for gunners evolved along a different path. By and large, the empirical mode of prediction became institutionalized. Proving grounds became locations endowed with considerable amounts of resources for not only measurement and testing but also computation. Giant ballistic pendulums or electric chronographs were designed to measure muzzle speed and air resistance even more accurately.[44] The goal of ballistics included being able to express air resistance as a function of speed obtained from experimental firings. In France, the Gâvre Commission pursued this goal, as David Aubin (2017) has documented so well.[45] This work involved mathematization in a particular sense—namely, designing computational schemes that navigate between different conflicting goals and conditions. Including an account of air resistance and the Magnus effect made them scientific in the first place. But the theoretical backbone had to remain flexible enough to be adaptable to the growing amount of data. At the same time, whatever approximation scheme was introduced, it had to be efficient in terms of computational resources and time. Mathematicians such as Siacci (1892) or Moulton (cf. Gluchoff [2011]) aimed at mathematical methods that strike a balance.

Ballistics had started out as the first field of new sciences in which mathematization promised to serve prediction. The empirical and the rational mode of prediction slowly crystallized into a state of being in conflict with each other, but often in a progressive way. In the early twentieth century, however, prospects changed: approximation, computation, and economy posed new demands. Mathematization was not just about a refined account of physical complications. It was more about constituting a new subject matter for mathematical methods: numerical approximation and efficient computation. However, enlarging the toolbox of mathematics by adding stepwise and iterative algorithms turned computation into a bottleneck. Consequently, the empirical culture of prediction started to organize computation on a large scale.[46]

The story of the ENIAC computer would fit here.[47] A World War II development, the original impetus for a large computing machine was to aid in calculating ballistics tables for use by the US Army. The project began as a collaboration between the US Army Ballistics Research Laboratory at the Aberdeen (Maryland) Army Proving Ground and the Engineering School at the University of Pennsylvania. Firing tables were necessary to allow gunners to aim large ordnance properly; and during the war, large numbers of new designs all had to have accurate firing tables to be used in the field. Originally, teams of female human computers were assembled to make the necessary repetitive calculations, but this was slow (not due to the lack of skills of the mathematicians but rather to the scale of the task) and, at times, depended on other calculation machines, such as the differential analyzer, that were also in short supply. The complication to this story is that by the time the computer was completed in 1946, the war had ended; ENIAC did nothing to speed up the calculation of ballistics tables, and even before the end of the war, it had been repurposed to aim at calculations for the Manhattan Project's atomic weapons through the intervention of John von Neumann.

Here we leave the history of ballistics. It has served to introduce two different modes of mathematical prediction. As indicated previously, from the late nineteenth century onward, ballistics would lead to a new, numerical mode of prediction that spans pre- and postcomputer time and flourishes in the mainframe culture of prediction (see chapters 4 to 6). The following chapter 3 on engineering epistemology will analyze how mechanical engineering in the late nineteenth century wanted to hybridize both modes of prediction, but had to struggle with methodological, disciplinary, and institutional conditions.

3 Engineering Knowledge, Autonomy, and Mathematics

How do engineers and scientists use mathematical tools to attain knowledge about the world, especially about the human-built world? Any answer to this question touches on deep philosophical assumptions about prediction, explanation, and rationality. This chapter takes a close look at the late nineteenth century in an effort to learn about mathematical tool use and, simultaneously, engineering knowledge.

The late nineteenth century stands out as a time when engineering disciplines underwent formative changes.[1] Layton argues that industry needed engineering *science* because the new technology demanded a more scientific approach.[2] Layton (1986) might be right when he ponders: "The transition from traditional rule-of-thumb methods to scientific rationality constitutes a change as momentous in its long-term implications as the industrial revolution itself" (3). However, what is meant by scientific rationality here? This concept is a moving target, undergoing changes in line with the changes in engineering and its contested intellectual positioning.

In a way, this calls for such an examination as the one that follows. It is easy to perform an investigation when it is guided by well-established preconceptions. In such a case, checking such assumptions against historical cases becomes a difficult but noble task. We believe that wanting to learn about history and philosophy at the same time is one of the best ambitions to have. We show how engineering synthesized a culture of prediction. This is a hybrid culture that centers around a mode of prediction quite akin to the empirical mode introduced in chapter 2—and partly in opposition to a rational culture in science. Thus, equating mathematization with a strengthening of the rational mode of prediction would be just that kind of mistake we want to avoid. Detailed studies such as the one in this chapter are a way to illuminate the dynamics and plurality of math-based predictions. They shed new light on the epistemology of engineering. Engineers introduced and advocated a *hybrid mode of prediction* that combines both rational and empirical modes. In fact, the hybrid concept of mathematization plays a key role in arguing for the autonomy of engineering knowledge. Although the rational and empirical modes were older and already firmly established, their hybridization was a competitive innovation that shows how modes of prediction, mathematization, and technology coevolved. Consequently, the emerging highly dynamic picture of math-based predictions has implications for high-level questions about our knowledge, especially about the perceived rationality of mathematization.

This chapter chooses Robert Thurston as its starting point. He was arguably the most prominent engineer in the United States after he reorganized Cornell's Sibley College in the late 1880s. Seminal for our purposes is Thurston's quarrel with physicist Henry A. Rowland on the relationship between science and engineering. Against this background, we analyze the Anti-Math Movement [*Anti-Mathematische Bewegung*] in German engineering that ignited in the 1890s in the aftermath of the 1893 World's Columbian Exposition in Chicago.[3] We point out that this Anti-Math Movement, contrary to the name it has acquired, was not, in fact, directed against mathematization. Rather, it unfolded in a conflict between two modes of prediction: the empirical and the rational. A new hybrid culture of prediction was emerging, accompanied by controversies on mathematization and engineering knowledge, as the fight between Alois Riedler and Felix Klein reveals. The new culture is exemplified here by the work of the engineer Carl Bach, one of the most prolific and influential mechanical engineers of this time. He is known as the founder of the Materials Research and Testing Station in Stuttgart in 1884. We analyze his role in the "Movement"[4] and examine a series of examples of how Bach combined experimental and mathematical work, thus promoting a hybrid mode of prediction. For him, systematic experimentation and mathematical modeling should be mutually interdependent in order to deliver usable and relevant predictions. Bach was by no means singular in this respect. On the contrary, his work was part of a wave; he was one of several pioneers who came up with a similar concept of hybrid prediction around the same time such as Oliver Heaviside (United Kingdom) and Charles Steinmetz (United States). The final section elucidates how the new culture of prediction challenged the given understanding of scientific rationality. We identify a characteristic line of conflict—namely, the debates over whether the hybrid mode establishes rules or laws. This conflict reveals the tensions between prediction, explanation, and rationality that are so characteristic of engineering epistemology.[5]

3.1 Putting Engineering on the Map: Henry Rowland versus Robert Thurston

In the late nineteenth century, both physics and engineering were in a phase of substantial growth and searching for a professional identity. A key controversy was whether and how pure science, applied science, and engineering should be demarcated from one another. Was there continuity, or even hierarchy?[6]

Henry Augustus Rowland (1848–1901), shown in figure 3.1, was an iconic proponent of the standpoint that would put pure science highest and engineering implicitly at the low end. Rowland, a physics professor at Johns Hopkins University, was America's foremost physicist of this time. He had studied with Helmholtz in Germany and remained a lifelong admirer of the German university system. From 1899 to 1901, he was the first president of the American Physical Society. Rowland became world famous for his handling of high-precision instruments.[7] His diffraction gratings resolved 600 lines per mm and increased the accuracy of spectroscopy by

Figure 3.1
Henry Augustus Rowland (1848–1901). Courtesy: Wikimedia commons.

a factor of ten. These "Rowland gratings" formed the worldwide standard in spectroscopy for decades. Speaking as vice president of the physics section of the American Association for the Advancement of Science (AAAS) in 1883, Rowland delivered "A Plea for Pure Science" (Rowland 1883) in which he distinguishes pure from applied science. Whereas applied science has practical value, according to Rowland, it rests on pure science—which was still underdeveloped in the United States.[8]

Rowland was a beacon of and an activist for the rational mode of prediction. He describes science as a logico-mathematical system in which laws discovered by pure science allow scientists to derive predictions from them. Once the theory is there, what remains to be done is to exercise one's skills of derivation. Rowland expressed his position as a challenge to engineers—for example, at the 1884 Electrical Conference in Philadelphia he declared that

> every law of electricity necessary to be known is already known; it is only a question of the brain that has the power to evolve the perfect machine, and when we say that theory does not agree with practice, it means that we have not got brains enough to apply the theory to the facts and get at the result. (Rowland [1886, 111], cited in Kline [1995, 200])

Rowland's claim is a strong and audacious one that leaves little room for interpretation. Engineering is cast into a role of depending—mathematically and therefore in general—on pure science.[9] Remarkably, Rowland does not mention potential problems on the mathematical side; he does not reason about what tools are needed or available, leaving it with the somewhat sophomoric "brains enough." We find this blind spot to be pivotal. Our analysis leads us to the opposite claim: making math-based predictions relevant for engineering required a more empirical, hybrid mode of prediction markedly different from the rational mode. This chapter examines how engineers developed this mode. Thus, rather than being the grounds for the dependence of engineering, it is mathematics that strongly contributes to its autonomy.

Rowland's plea did not go unchallenged by engineers. Robert Thurston (1839–1903, shown in figure 3.2), the most prominent mechanical engineer of this time, took up the gauntlet and used his 1884 address to the section for mechanical engineering of AAAS in Philadelphia to outline the "Mission of Science" as a sophisticated alternative to Rowland's plea. Thurston was a specialist in iron and steel as well as steam engines.[10] He began his career as an assistant professor at the US Naval Academy in Annapolis; and from

Figure 3.2
Robert Thurston (1839–1903). Courtesy: Wikimedia commons.

1871 onward, he headed mechanical engineering at the Stevens Institute. Thurston not only adhered to a school culture based on a European model but also emphasized that experimental facilities were a necessary part of engineering science. He started an experimental engineering laboratory at Stevens[11] and, in the same year, conducted a series of experiments on steam boilers on behalf of a committee of the American Institute in which, for the first time, all losses of heat, all steam generated, and the quantity of water entrained by the steam were measured with high precision. Based on these observations, Thurston built a quantitative, predictive model that helped standardize and regulate boiler technology. This commissioned work presents an early example of how engineering would increase predictive power by combining mathematical and experimental approaches.[12]

The year 1884 was special for Thurston, and not just because of the Rowland challenge that led him to defend the status of engineering science. The year brought a new offer for Thurston because Cornell had set up a committee that was looking for a new head for Sibley College charged with reorganizing the entire school of engineering.[13] Thurston began as an outside advisor but eventually accepted the post himself. His "Mission of Science" address provided a programmatic meaning for what he planned at Cornell. "The mission of science is the promotion of the welfare, material, and spiritual, physical and intellectual, of the human race" (Thurston 1884, 231). This general approach embedded engineering in a scientific *system* while reserving a special role for it as the discipline able to cash in the promise of the mission. With that role granted, engineering can transform smoothly from art to science: "In the past the arts have led; in the future we shall see science leading and directing every development of the arts" (Thurston 1884, 233). It is the utilitarian values that elevate science to "applied science"—a notion Thurston uses for engineering science.

Thurston continues his mission talk with a particular section on engineering methodology and epistemology in which he advertises his own approach—that is, first developing a precise basis of empirical measurements under varying conditions and then distilling predictive (mathematically formulated) principles out of these data. If development had been slow in mechanical engineering, it was because this approach had not yet taken root.[14]

> But it is only within the past few years that the conditions modifying the value of these materials, as applied in engineering, have been carefully and critically studied by the light of experimental investigation. The effect of heat on strength and elasticity; the alteration of structure produced by vibration. (241)

Many more problems of this sort remained, including mechanical properties of materials, the effect of temperature on the materials, the corrosion of steam boilers, the value of lubricants, and the heating power of fuels.

Thurston is clear: Rowland's position does not hold water because application is *not* a question of brainpower. Rather, it is a question of combination—Ron Kline (1995) speaks of "hybrid theory" (202–203). In the cases in which engineering is interested, prediction is beyond the power of math alone, requiring experimental testing and guidance. The mathematically formulated theory or model must be modified and amended by empirical data. How this can be done is again a question of mathematical modeling. In other words, Thurston, in effect, involves elements of both of the rational and empirical modes of prediction. Only the hybrid promises the predictive power engineers were seeking. In this way, engineering would be an applied science *without* depending on pure science.[15]

Going beyond engineering epistemology, Thurston (1884) addresses education and institutions and does not fall short of promoting a nationwide architecture—"schools of science in every city, colleges of science in every state"—in full contrast to Rowland's elitist ambitions (245). In other words, Thurston is promoting a culture of prediction that includes the hybrid mode of prediction as a component. Thurston fully convinced the Cornell committee and especially President White, who saw Thurston's work at Sibley College as an epochal turning point: "So began a new era for Sibley College, for Cornell University, and indeed, for the country (. . .) Professor Thurston's coming put an end to all divided counsels and began a new and better order of things" (President White in 1915, cited in Durand [1939, 549]).

Thurston wanted to perfect the model of a pure technical school that would lead into research (arguably an elitist ambition, too). He raised entrance conditions, in this way delegating elementary work to other institutions. The curriculum included science, higher mathematics,[16] and language in the first two years—creating the expertise among students that they would need to work as engineering researchers in the combined experimental–mathematical context. The last two years covered specialized topics such as steam engineering. Some shop experience and training in drafting were given alongside this (see Calvert 1967, 97).

Apparently, Thurston was an extraordinarily efficient organizer. In 1890, after only about five years, he wrote to Cornell's president that his mission was now accomplished.[17] That leads us to the 1890s, when Thurston was once more at a threshold in his career. Cornell was now the leading place to study engineering in the United States, and Thurston was disseminating nationally and internationally.[18] He turned into an ambassador for a hybrid culture of prediction.

The most prominent stage was at the 1893 World's Columbian Exposition in Chicago. This exposition triggered a flurry of reports and programmatic statements in which US engineers debated the status of engineering in the United States and the status of US engineering worldwide. Here, we are particularly interested in how claims for a new culture of prediction reverberated almost instantly in Germany. A quick glimpse at contemporary journals and newspapers suffices to show that both Americans and Germans felt that they were at the height of scientific and engineering accomplishments. Germany was a widely acknowledged model for university science and still a role model for the next generation of US scientists. However, when it came to the economic and academic success of engineering, Americans felt on par and even in the lead because of their experimental facilities for engineering. And they wanted the world to know.[19]

3.2 The 1890s Anti-Math Movement in German Engineering

From here, the focus of our investigation shifts to Germany. What happened there immediately afterward documents both how central and how controversial the hybrid culture of prediction was for the developing identity of engineering science. The meetings at the 1893 Exposition started a chain of events that sparked a fierce controversy known as the Anti-Math Movement of German Engineering [*Anti-Mathematikerbewegung der Technikwissenschaften*]. We need to backtrack a bit because telling this story demands careful exposition. During the later part of the nineteenth century, Germany saw a rapid rise in science and industry.[20] German engineers struggled to find, or define, their place.[21] There was an intense debate around engineering and its relationship to professionalization, science, education, and industry.[22]

Regarding education in technical institutes, the mid-nineteenth century was dominated by professors of engineering who maintained a theoretical orientation in engineering and emulated the path of mathematized natural science. The leading persons were Ferdinand Redtenbacher (1809–1863), Franz Grashof (1826–1893)—both at Karlsruhe Polytechnic School—and Franz Reuleaux (1829–1905) in Berlin (since 1879, the Königlich Technische Hochschule/Royal Technical Institute Charlottenburg).[23] Consequently, the education of engineers was oriented toward mathematics with the goal of rendering questions of design and construction calculable—without caring much about whether the promised practical utility would actually be delivered.[24]

Teaching mathematics had changed significantly over the course of the nineteenth century. Whereas Gauss had given courses on calculation techniques and geodesy in the early 1800s, engineering students in 1870 would learn more abstract and "rigorous" math. One illustration is how mathematicians dealt with continuous functions. Continuity had long been a prime example of an intuitively graspable and physically relevant concept. Euler, for instance, defined a continuous function as one whose graph can be drawn without raising the pencil. In the mid-nineteenth century, however, functions were conceived of as arbitrary mappings, and this accordingly called for a more technical and abstract approach to continuity. Consequently, universities produced teaching professors with increasingly abstract conceptions of mathematics.[25]

In short, by the 1870s, engineering was trapped between two standpoints—a strong faction of industry-oriented engineers on the one side, who favored usability over academic status, and science-oriented professors on the other side, who took mathematics and physics as their role model.[26]

Against this background, we shall take another look at the 1893 Exposition in Chicago, this time from the German side. Alois Riedler, mechanical engineering professor at Berlin, reported from the 1893 Exposition in Chicago and, even more importantly, from his trip to visit American technical institutes. In a study commissioned by the German Federal Ministry of Education (Riedler 1893). Much to the satisfaction of politicians and colleagues alike, Riedler confirmed that German products were now competitive. But Riedler's goals were bigger: he used the comparison with the United States to see which lessons German engineers needed to learn. He made two major points: first, he observed that the US system was based on much less mathematical education than the German. Second, Riedler identified the well-equipped laboratories, like the one at Cornell, as a crucial success factor.[27]

In the aftermath of the Chicago Exposition, regional German engineering associations discussed Riedler's report and its possible recommendations for engineering education. Adolf Ernst, professor at Stuttgart, was commissioned by the Verein Deutscher Ingenieure (VDI) to synthesize these discussions into a report that he presented in 1894 to the general assembly. This report, titled "Mechanical engineering laboratories" [*Maschinenbaulaboratorien*], demanded engineering laboratories through an expensive upgrading of technical institutes (THs).[28] In agreement with Riedler, Ernst (1894, p. 1354) diagnosed that the education of engineers was suffering from a lack of experimental facilities and that the time assigned to laboratory instruction should be taken from the mathematical part of training.

One factor that aggravated the perceived urgency of the situation was presented by Felix Klein, the renowned mathematician and influential science organizer. He had come forward with plans to establish a research laboratory and teaching institute on "technical physics"—that is, something akin to engineering science—at the university in Göttingen. This was not a mere coincidence. Klein had also been in Chicago: as a main speaker at the Mathematics Congress.[29] He took the lessons about experimental facilities as an opportunity to promote applied mathematics. This proposed institute for "technical physics" [*Technische Physik*] would tie together applied mathematics and mechanical engineering. This plan met strong resistance from the side of engineers because the institute would be located at a university without an engineering school. The engineers perceived this as an attempt to strengthen research related to engineering science, but also to funnel resources away from technical schools and toward traditional universities. Because Klein stepped forward at exactly the time when the VDI was chewing over the Ernst–Riedler reports, the need for a reaction was considered urgent, and the VDI felt it had to support Ernst–Riedler and rebuke Klein—that is, perform the somewhat tricky feat of confirming its scientific status, claiming research facilities, but stopping research from being absorbed by scientists at universities.

The laboratory movement quickly escalated into the *Anti-Math Movement* [*Antimathematikerbewegung*].[30] Most historical studies draw an overly simplified picture of this movement. Of course, there was an influential (and loud) faction that was quite literally against math. There was a second faction, however, that promoted a hybrid concept of mathematization that integrated experimentation with mathematization. This second faction managed to infiltrate the "Aachen resolutions" from 1895 in an important way.

The 1895 Aachen Crash and Subsequent Recuperation

The governing board of the VDI had set up a commission, led by von Borries, to synthesize what recommendations the VDI should issue following the reports of Riedler and Ernst. The commission came up with a text (von Borries 1895) that confirmed Ernst's stance, claiming furthermore that the VDI should support the *Technische Hochschulen* (THs) in this agenda. The text was acclaimed by the 36th general assembly [*Hauptversammlung VDI*] of German engineers in the town of Aachen and became known as the Aachen resolutions [*Aachener Beschlüsse*] of 1895.

Here is a summary of the nine points: (1) THs should provide not only a good education for average engineers but also opportunities for research;

(2) engineering laboratories are needed for (3) practical exercises during education and for research; (4) courses in laboratories should become obligatory, whereas (5) other parts of the curriculum should be shortened; (6) teaching in auxiliary sciences should be oriented toward understanding engineering sciences. In particular, abstract mathematical methods should be restrained, though not at the cost of an understanding of engineering. Use of math in applications might make it easier to master it as a tool; (7) point 1 renders it necessary to create facilities for physical-technical education in theoretical and experimental respects going above and beyond the general curriculum; (8) final exams should be geared toward general technology and less toward examinations for state service; and (9) teaching in laboratories cannot replace gaining practical experience in the shop. The latter should last one year at least.

Most scholars take the Aachen resolutions as documenting the revolt of engineers and their desire for emancipation from the universities' monopoly on doing science. They allegedly expressed their wish to reduce mathematics and resort to experimental laboratories instead.[31] Going through the points, however, does not strongly indicate why this should be proof of an Anti-Math Movement. In fact, the wording of the resolution intentionally covers over the strongest oppositions. A last-minute revision had happened. Taking this situation into account, the Aachen resolutions should be read as a diplomatic feat, initiating an agreement between all stakeholders.

This story was revealed only as late as 1998 by Puchta (formerly Hensel) when she retrieved Bach's letters in the archive of the TU Chemnitz. Peters, in 1895 the acting secretary of the VDI (the powerful position newly established in the VDI's reorganization), was alarmed that the assembly, eager to show strength, could vote for a resolution that would depict engineering as something separate and independent from science, and that it would cut off all institutional and educational ties. Peters saw this as counteracting the autonomy of engineering science—and he knew he had company in this belief among those engineers who had already worked toward establishing engineering as a science, such as Linde (Munich) and Bach (Stuttgart). Peters invited Bach and Linde as additional leading authors in formulating the text of the resolutions.[32] Possibly, Bach is held as a ringleader in the Anti-Math Movement because he is named as a lead author and he gave the most substantial contribution to the discussion of the Aachen resolutions at the general assembly.[33]

This brings us to the second reason—namely, the content of the resolutions. Bach introduced crucial last-minute changes to the already drafted resolutions. In fact, Bach was the one who added point 1 and hence brought in a new line of thought—that is, that technical institutes should also do cutting-edge research and that this would require a two-pronged approach consisting of an experimental *and* a theoretical part. For the first, laboratories are necessary; for the second, theoretical, mathematical modeling is key. Both have to come together. Thus, Bach captured the force of the laboratory movement and, at the same time, strengthened the role of mathematics. To make sure his point was accepted, Bach defended it in front of the general assembly.

In the discussion at the VDI assembly, Bach (1895) underlined that an average engineer does not need as much abstract math as is now obligatory—thereby appeasing the anti-math faction (see also Bach [1896]). Bach explained that point 1 of the resolutions was targeted at those engineers who would turn toward research in engineering science. Because engineering research is dependent on mastery of mathematics as a tool, Bach planted implicit support for mathematics into the resolution—for which he got unanimous support from the assembly. He went even further, brought Klein to Aachen in 1895,[34] and promoted an agreement with Klein's initiative, the *Aachener Frieden* [Aachen peace]. Bach wanted to make sure that resources for research in engineering would go to technical institutes, not universities. Klein's institute of "technical physics" would be acceptable if Klein were to make it clear that he did not address (research-gifted) engineers but restricted the audience to physicists and mathematicians interested in engineering. In this way, one could even join forces (though this did not happen officially). Both men proposed a hybrid between applied math and mechanical engineering, though with a markedly different understanding of the math part.

The dust did not immediately settle after Bach's initiative. He had designed a mediating resolution but could not prevent more militant standpoints from being articulated. The following half decade saw an extension and partly a deepening of controversies around engineering and its scientific, educational, and institutional organization. In the aftermath of Aachen, engineers and mathematicians engaged in a mass confrontation of sorts that eventually justifies the name "Anti-Math Movement." Because voices calling for the downgrading of mathematics did not vanish, in 1897, all thirty-three professors of mathematics and mechanics at THs signed the

"Darmstadt resolution" that opposed the status of being an auxiliary science while maintaining the foundational status of mathematics.[35] A direct answer by fifty-seven engineering professors insisted on the merely auxiliary role. The 1897 VDI assembly in Kassel came near to rolling up Bach's plan. It proposed a vote on banning higher math entirely from engineering education—the ban was voted down by a narrow majority.

Over the next decade, the Aachen peace prevailed. A group of leading engineers and mathematicians, with Bach among them, went on to advertise the mutual benefits of hybridization.[36] Engineering laboratories were created at all THs. However, training did not change to a very great extent. Reports suggest that students still did not receive hands-on lessons. Instead, these laboratories strengthened the research profile of engineering. In terms of academic status, the THs obtained the right to award doctorates.[37] Klein's agenda was also successful. The Göttingen institute was founded with engineer Ludwig Prandtl, who had been Föppl's assistant in Munich, and mathematician Carl Runge, who accepted Germany's first chair for applied mathematics.[38] However, in this agreement, the hybrid concept of mathematization worked as an umbrella term that hid crucial differences regarding engineering epistemology. Appreciating the emerging new (hybrid) culture of prediction requires a closer look that will resolve these differences.

We shall now discuss and compare three of the main actors: Riedler, Klein, and Bach. The first two were true opponents. We shall analyze their controversy in the following subsection (3.3). Bach, the third actor, was not only the go-to mediator and bridge builder but also pioneered a hybrid culture of prediction (see section 3.4).

3.3 Riedler versus Klein: Different Takes on Mathematization and Engineering

Alois Riedler (1850–1936, shown in figure 3.3) and Felix Klein (1849–1925, figure 3.4) were both leaders in their respective disciplines; both were convinced that new technologies would call for a new conception of mathematization; both were pushing new initiatives in education; and both came to opposite conclusions. The role of mathematics in engineering science marks the core of their disagreement. Whereas Klein defended the viewpoint that mathematics is a foundational science and, consequently, that engineering is scientific to the degree it is subsumed under mathematized applied

Figure 3.3
Felix Klein (1849–1925). Courtesy: Wikimedia commons.

Figure 3.4
Alois Riedler (1850–1936). Courtesy: Wikimedia commons.

science, Riedler maintained the opposite: mathematization is a good and useful thing only insofar as it fosters the autonomy of engineering.

Felix Klein was a geometer who studied in Bonn and Berlin and held professorships in Erlangen, TH München, Leipzig, and (from 1886 onward) Göttingen. In mathematics, he is still known for his 1872 "Erlangen program" to characterize spaces based on geometry and group theory. Later in his career, he organized the rise of Göttingen to a world-leading center of mathematics and mathematical physics (in the early twentieth century). Klein was deeply involved in teaching and in reasoning about mathematical education. The nineteenth century had seen the invention of pure mathematics and its rise to the forefront in nearly all quarters of math research. Klein fully appreciated this fact that opened up many new avenues for the discipline (not least pursued in Göttingen), but he also acknowledged the relevance of applied mathematics because the prominence of mathematics (as a discipline in the university) rested on its significance for the sciences. Klein realized that technology was posing a new challenge: it demanded accurate predictions in increasingly complex circumstances in which traditional rational mechanics turned out to be rather helpless.

Klein perceived this as an opportunity to increase the relevance of mathematics.[39] Applied mathematics should be developed to serve as a backbone for applied science, including mechanical engineering. According to Klein, pure and applied share the basic rationality—natural laws are described as mathematical functions—but in the applied case, additional conditions of technology have to be taken into account. How one should do this, Klein held, called for a mathematical research program. When proposing his vision, Klein usually addressed an audience of mathematicians and physicists where he faced opposition to his appreciation of applied science (cf. Manegold [1970]). Thus, Klein's strategy was to embed applied mathematics into a broader appreciation of pure science and mathematics.

At the 1893 World's Columbian Exposition in Chicago, Klein was a main speaker at the international mathematics congress. There he praised the value of applied mathematics and its use in applied and technical sciences (Klein [1894], sixth conference). Klein learned in Chicago how much American engineers were convinced that experimental facilities are key for engineering science. Klein realized he could ride a wave here, and he quickly developed an agenda advocating for a new type of research institute that should open up university research to the demands of technology and engineers. He wrote a memorandum on a new institute for "technical physics" in spring 1895 that attracted immediate attention as well as a negative reaction from the side of engineers as described previously. In a series of further talks, both to mathematical and engineering audiences, Klein tried to show how his plans were in the good interest of every party, including engineers (Klein 1895, 1896a, 1896b, 1898). These talks were met with criticism, often expressed in direct commentaries on the publication. Klein's talk to engineers on "Demand of engineers and education of mathematics teachers," for example, appeared in *ZVDI* (the publication journal of German engineers) in 1896 with harsh comments by Riedler. Where Klein argued for integration into modern mathematics, Riedler called for separation.

Klein not only designed a new institution but also argued about mathematics itself—in particular, about the mathematization of approximation techniques. Klein sought to make the case for the applied side by subsuming it under the rational mode of prediction (cf. chapter 2 of this book). Throughout his work, he was "always underlining the relationship between the exactness of the idealized concepts and the approximations to be considered in application" (Menghini 2019, 181). Klein described how a new technical mechanics [*technische Mechanik*] should adhere to the rational ideal but also differ from classic (rational) mechanics in four points: (1) problems must be less idealized (e.g., retain effects of friction); (2) some aspects are data-driven (because laws do not cover complex phenomena completely); (3) approximation (not idealization) is the goal for mathematics; and finally, (4) graphical methods (instead of analytical ones) are more usable for engineers (Klein 1900b). In a 1902 lecture course, Klein undertook to detail the relationship between "Precision Mathematics and Approximation Mathematics,"[40] including geodesy, drawing, and (as a central piece) the "approximated representation of functions," with the intention of basing predictions on numerical methods.

In this way, Klein retained the rational mode and amended it with an approximation component.[41] Overall, Klein stressed the unity of science and a sort of top-down mathematization. In his view, this would elevate technical mechanics to a science and connect it to the rational mode of prediction. Klein did not gain approval from the side of engineers. Riedler, Klein's most outspoken critic, opposed the unity of science and top-down mathematization, breaking it down into an autonomy of (engineering) science and bottom-up mathematization.

Alois Riedler (1850–1936) studied machine construction in Graz and Brünn before holding professorships at TH München, TH Aachen, and (since 1888) TH Berlin Charlottenburg where he was also appointed president (*Rektor*) from 1899–1900. He was a specialist in pumping machines and fast-moving motors. Riedler was also a prolific publisher on engineering education, starting with the report on American technical institutes from his visit in 1893 (he particularly praised Cornell, MIT, and Stevens) and including his monograph (1895) that he tailored for the Aachen assembly.

Although Riedler is rightly seen as the ringleader of the Anti-Math Movement, his stance toward math was not plainly "anti" but deserves a more nuanced look. He saw the importance of mathematization but had in mind a specific type of mathematics, namely graphical methods, geometry, and drawing. Apparently, Riedler rarely backed away from a controversy. He attacked Reuleaux's (his colleague in Berlin) vision of a theoretical kinematics (Reuleaux's pride) and threw it out of the curriculum.[42] At the same time, in every publication, Riedler (1896) hammered home that mathematics "is an indispensable foundational tool, but not itself the foundation" [*unerlaessliches Grundwerkzeug, aber nicht Grundlage selbst*][43] (305). Conversely, engineering science is not defined by mathematics but rather by the way it uses mathematical tools to solve engineering problems.

In fact, he scoffed at engineers who did not use mathematical tools for design. For instance, Ziese (at the then famous Schichau shipyard) had designed a fast ship competing for the "Blue Ribbon" (fastest crossing of the Atlantic Ocean) as a scale-up of a fast torpedo boat. More academically trained engineers with some formal mathematical knowledge could have told him, Riedler said, that this would not work. And indeed, the boat was fit only for scrap when it arrived in America.

For Riedler, connecting to some rational scientific structure was irrelevant. Instead, he underlined the flexibility of mathematics. One could use coefficients and parameters whose determination was left open for experimental measurement. Such empirical ingredients were no threat to rationality[44]—on the contrary, for Riedler, rationality lay in the efficient determination of coefficients. The engineer would formulate a problem whose efficient solution required mathematical and scientific means. Therefore, technical knowledge is on a higher level than mathematical (scientific) knowledge (Riedler 1898, 5). Here, Riedler comes close to Thurston's position in the "mission" discussed previously.

Thus, for Riedler, mathematization amounts to seeing engineering and science from a methodological point of view[45]—that is, as a tool for tackling engineering problems—not as an entry point into the rational structure of nature. Mathematics is a flexible tool without ontological import. The most important goal for engineering is designing a device or organizing a system of devices so that they are "efficient in industrial practice" [*betriebsbrauchbar*[46]]. Riedler saw that the complex conditions of modern technology required experimentation (in engineering laboratories) as part of finding an adequate formulation of problems. "Many are able to build machines; often practical experience is sufficient; but building machines and arranging them to the greatest practical benefit, this is the task of scientifically minded engineers" (Riedler 1898, 115).[47]

In sum, Riedler proposed mathematization but along a line very different from Klein: not as a link to the broader theoretical structures of science but as a flexible tool that served the autonomy of the technical (or engineering) sciences, and that could be called bottom-up mathematization because it follows the needs of the problems at hand. As with Klein, mathematics is useful in creating predictions, but Riedler's stance is much more like the empirical mode of prediction discussed in chapter 2.

3.4 Carl J. Bach: Building a Hybrid Culture of Prediction

Carl Julius Bach (1847–1931) was one of Germany's leading mechanical engineers at the time. We have come to know him as a main actor in the Anti-Math Movement who pulled strings behind the scenes, thus mediating the controversies not only within the engineering community but also between engineers and scientists.[48] What makes him particularly interesting in our context is that he proposed a hybrid culture of prediction as the way for engineering science to proceed. In fact, he implemented it. Bach pushed for experimentation *and* mathematization in a way that combines the older rational and empirical modes of prediction discussed in chapter 2. We do not claim that this made Bach unique. Obviously, Riedler and Klein debated on, or fought for, their version of this mode. Approaches in engineering research tended to converge toward this mode both in the United States (at Stevens, Cornell, and MIT among other locations) and in Germany (examples are the Munich and Göttingen institutes of technical physics). However, Bach articulated the hybrid mode in an influential way for the engineering profession. Moreover, we are also not claiming that Bach was a typical case. All actors discussed here, from Thurston to Riedler, Klein, and Bach, were pioneers who led their profession but were (almost per definition) exceptions in their time.[49]

Bach started as a worker in the steel and then steam engine industry without any academic education. He began his academic education in 1866 with a stipend at the Polytechnicum Dresden interspersed with periods of practical work. In 1872, he studied at the Karlsruhe Polytechnicum (TH), where he obtained his diploma.[50] Bach then worked as an engineer in Germany, Austria, and England before becoming director of the "Lausitzer Machine Factory," Bautzen (Germany) in 1876. In 1878, he was awarded a professorship at TH Stuttgart. He did groundbreaking work on the strength of materials [*Festigkeitsforschung*] after setting up an experimental facility for this task. The *Materialprüfungsanstalt* [Institute for testing materials] opened in 1884.[51]

Figure 3.5
Carl Julius Bach (1847–1931). Courtesy: Wikimedia commons.

Bach held mathematics in high esteem. In his autobiography (1926), he reported that he was the most skilled student in mathematics at Dresden. Proud that he had outwitted an older student who had held this position and later became a mathematics professor, Bach (1926) recalled that he had to give a preliminary course in mathematics at Stuttgart (20). After the first semester, even students who had already heard the regular course in math attended Bach's course because of its intellectual accessibility. In a passage on mathematics that appeared in the foreword to the third edition of his book on elasticity, published directly after the 1895 Aachen incident, Bach (1920) articulated his position combining a leading role for experiments with a necessary role for mathematics: "In the engineering sciences, securing and enlarging empirical foundations has to assume priority. In this, mathematics will not only be an extraordinarily valuable tool, but will present the tool without which any deeper understanding would remain unattainable" (preface to third edition [1898, ix]).

In the 1890s, Bach was one of the most active and influential engineers in Germany. New materials (often varieties of steel) and more complex machines had made systematic knowledge about the strength of materials a bottleneck for mechanical engineering design. Bach made a name for himself by addressing this bottleneck. He could talk to both practitioners and the academic elite (similar to Thurston). Part of Bach's high reputation came from two very popular books. The first, *Elastizität und Festigkeit* [Elasticity and strength of materials] appeared in 1889 and saw nine further editions up to 1924[52]; the second, *Maschinen-Elemente* [Elements of machines], applied the findings on the strength of materials to the design of machine components. The first edition came out in 1896 and the thirteenth and final edition in 1922.[53] His books were a tremendous success with an entire generation of engineers. Written in an accessible style, they showcased his hybrid approach to prediction—and thus made a case for the autonomy of engineering. In 1918, Bach—from then on "von Bach"—became the first technical engineer to be granted the official title "Excellency."[54]

From here onward, we shall examine how Bach combined experimentation and mathematics to obtain predictions. Bach presented a confident synthesis of various cases in *Elastizität und Festigkeit*. In instructive forewords (keeping the forewords from earlier editions), Bach reflects on his own approach. His work "starts from the assumption that the primary concern is knowledge about the factual behavior of materials" (Bach [1920], preface to first edition [1889, iii]). Bach's typical approach is to investigate where existing mathematical rules in engineering in fact either made good predictions or failed to do so. First, systematic experimentation is needed—for instance, to determine strain in various materials and situations. This typically results in tables that collect these data for certain forms and materials. Second, these data are used to test existing (theoretical) mathematical rules—usually predictions in the rational mode. If the existing rules are not sufficiently adequate,[55] new rules are sought via modification of the mathematical form. This two-pronged procedure hybridizes empirical fit, theoretical derivation, and skillful modification. Mathematization then is about having a flexible tool for creating predictive rules. Such rules have to fulfill two counteracting conditions: they have to be sufficiently general to be useful in new designs, and they have to be empirically adequate—give correct predictions—in practically relevant circumstances; and this, in turn, presupposes that they are tractable in practical cases.

Bach (1889) criticizes the viewpoint that engineering sciences will be further advanced via mathematical derivations from taken-for-granted theories and laws (452). An illustrative example is Bach's short piece on "Common mistakes in certain hydraulic calculations" (1891) in which he challenges the theoretical gurus of his time.[56] Bach questions the commonly accepted rules for calculation because they might be misleading in the case of pipes that have sudden changes in diameter. He documents his suspicion by experiments showing that the predicted slowdown of fluid velocity does not occur. Hence, these rules are disconfirmed by Bach's experiments. Bach goes on to analyze the mathematical derivation of the commonly accepted rules given by Reuleaux, Grashof, and Weisbach—the leading theoretical accounts of his time.[57] Bach is able to single out one presupposition that was used in all these theoretical treatments but is, in fact, at odds with experimental measurement: namely, they all wrongly assume an inelastic collision when the cross-section is broadened. Bach also recommends mathematization but warns against following some perceived rational logics of mathematics. Of course, idealizations are necessary in Bach's view as well, already because tractability is a necessary condition, but they have to retain contact with experience and quantitative values from practical situations. Thus, one needs experimental facilities and mathematical tools working together.

Another example is Bach's work on the resistance of rivet connections against sliding (1892). Using rivets was a popular design approach in building larger structures (see figure 3.6), but it had also seen a number of failures. According to the widely accepted theory of the time, the strength of the connection resulted from the strength of a rivet and the number of rivets—that is, the crucial parameters are the material and diameter of a rivet plus the number and location of rivets. Bach is not convinced and sets up systematic experiments that disprove this theory. He identifies the drag between riveted materials as the pivotal parameter. This shows that the strength of a construction depends not so much on the number of rivets but crucially on the carefulness and accuracy with which these rivets have been affixed—rivets that are not tight do not add to the drag. As a result, Bach can point out a major factor in design failures: "The usual way of calculating resistance against shear is likely to be the main reason for the insufficient strength of many iron constructions" (1892, p. 1141).

Hooke's Law and Bach's Law

Bach had to struggle in two ways with the character of the rules that would result from his strategy. First, dislodging long-hedged pillars of the discipline made it imperative to come up with replacements that performed better. Second, the newly established predictive rules could not claim the scientific authority of the established rational laws. The theory of elasticity is a main example for Bach's hybrid approach to prediction documented in his widely influential book on elasticity (1920, first edition 1889). The accepted theory of elasticity was based on Hooke's law. Bach (1926) intended to put

on firm ground those areas I worked in. However, I committed a sin in the eyes of some, for instance, when I turned the theory of elasticity from a discipline of the humanities into an empirical discipline. Before, the theory of elasticity started

Figure 3.6
Rivet construction. Courtesy: Wikimedia commons.

> from the theorem that tension and strain are proportional and then aspired to derive everything else by mathematical means. (32)

In the simple case of an (ideal) spring, elongation of a spring depends linearly on the mass (gravitational force) applied on the one end. The law then includes a constant factor that depends (only) on the spring.

$$\varepsilon = \alpha \times \sigma \qquad (*)$$

where α is a constant number[58] to be determined by experiment for each material. Roughly, the equation expresses the change of length per kilogram stress. For each material, there is a limit of proportionality—that is, the equation (*) becomes invalid for greater stress.[59] In other words, the theory of elasticity proceeds on the basic assumption that strain (elongation), epsilon, and stress, sigma, are proportional within certain boundaries.

Bach (1920) maintains that (*) is not a good basis for elasticity theory because it leads to false predictions: "This work shall show that it is not sufficient to assume stress and strain are proportional to each other, and, from this assumption, erect the building of elasticity theory by mathematical derivation" (preface to the first edition [1889, iv]). Instead, according to Bach, the design engineer "again and again" has to test the predictions and adapt the rules accordingly. Empirical experiments have the say, and predictive success or failure is the criterion that can show when mathematical considerations have used assumptions that simplify or idealize too much. According to Bach (1920), ignoring this condition has led to the slogan of a "contradiction between theory and practice" (ix). Bach holds that this contradiction disappears when mathematical modeling is embedded in a hybrid culture of prediction. "Science and technology have to go hand in hand" (ix).

Bach puts his own recommendations into practice and uses a variety of tests for different materials to show that α is *not* constant for many materials. No limit of proportionality is defined then because the material does not behave proportionally. However, the theory of elasticity presupposes proportionality as a basic fact.[60] Bach therefore changed the conceptual framing of elasticity when he modified the mathematical form, a modification that was called "Bach's law" (Equation 5a in Bach [1920]):

$$\varepsilon = \alpha \times \sigma^m. \qquad (**)$$

This equation contains two different coefficients (α and m) that both depend on the material. In a sense, the difference between m and 1 indicates nonproportionality ($m = 1$ would be Hooke's law). According to Bach, the power law (**) coincides well (for coefficients tuned to some specific material) with the data on all tested materials except for rubber and marble. This gives an apt impression of the sort of generality this law claimed. Elongation increases faster than tension for $m > 1$, which is the case for cast iron, copper, granite, concrete, and other materials. The opposite holds for, among others, leather and hemp ropes.

Bach notes that the two coefficients of (**) are very sensitive to changes in the composition and treatment of the material such as that occurring with calcination of steel. The coefficients are parameters that can be adapted to data but cannot be derived from theoretical considerations. Bach did not take his law to be a fundamental law of nature (or a nature of materials). Instead, he admitted that there might exist more elegant mathematical functions that attain a better fit.[61] For him, the main value of (**) consisted in being predictive about the relationship between stress and strain (elongation). Prediction would be impossible from the standpoint of Hooke's law. This kind of insight—being open-minded about the sort of mathematization that is appropriate—is exactly what was needed to make an accurate prediction.

Much later, in the preface to the seventh edition of 1917, Bach looks back to what had been achieved. When his book first appeared in 1889, the theory of elasticity and strength of materials had been considered an area of mathematical derivations alone, starting from the assumption of Hooke's law (proportionality of stress and strain). This law was conceived of as universal. By empirical testing and experimentation, Bach and others showed that most materials do not in fact obey this law. Nowadays, Bach resumed in 1917, the material plays as important a role as mathematics. He even concedes that the situation has reversed and that a surplus of empirical knowledge on properties of materials awaits mathematical–intellectual digestion.

Bach promoted mathematization bottom-up in a way that ran counter to the tradition of rational mechanics in which phenomena were identified more or less through their mathematical representation. His approach was also directed against an overly empirical approach that would replace mathematical modeling with gathering data. Thus, Bach promoted a hybrid mode of prediction that navigated between the rational and empirical modes.

Moreover, Bach was working toward a (hybrid) culture of prediction in which mathematics provided no more than one component. Another component is experimental facilities of sufficient power to detect inadequacies of the existing picture. But then one also needs people who are educated and organized in a way conducive to this hybrid approach. In short, engineers, with Bach among them, were building a hybrid culture of prediction.

3.5 Rules versus Laws: More on the Hybrid Approach

Although Bach was a pioneer of the hybrid approach, he was neither the only nor the first one. At around the same time (the 1890s), other engineers were devising similar approaches independently and in fields not limited to mechanical engineering. We have shown earlier that Bach ran into the problem of whether his "law" should count as a law or (only) as a rule.[62] Running into this problem is characteristic for pioneers of the hybrid mode of prediction. It does not just point out a moment of cultural resistance from the rational side, but it also has an important lesson to tell about engineering epistemology. Apart from questions of terminology, there is a deeply troubling lesson for engineering knowledge here. If predictive force results from partially abandoning the rational mode of prediction in favor of a patchwork of (hybrid) rules, this undercuts the perceived link between mathematization and the rationality of science.[63] Does engineering science uncover the rule of natural laws—or is it a systematic way to construct rules that work for predictions?

The common view held that the status of being a law is founded on generality, mathematical form, and derivation from more fundamental laws. However, rules such as Bach's law gain their generality and their predictive force exactly by dispensing with these factors. One controversial suggestion in the contemporary discussions was to categorize expressions such as Bach's law as being a technical law in contrast to a scientific law of nature proper. We suspect that there are more cases to be found and collected. What follows is a small sample of analogue cases. However, uncovering them requires a detailed look at how mathematical tools are used. Such a look is relatively rare in the historical literature.[64] We shall briefly introduce three cases, or rather vignettes, of the law-versus-rule debate.

Tetmajer's Law

Ludwig Tetmajer (1850–1905) was assistant to Carl Culmann, professor for statics at the Eidgenössisches Polytechnikum in Zurich (the later ETH), where Tetmajer himself became professor in 1878. Three years later, he advanced to a permanent professorship in building mechanics and, in parallel, became director of the newly founded *Festigkeitsprüfungsanstalt* [Institute for testing building materials]—from 1901 onward, the *Eidgenössische Materialprüfungs-Anstalt* (EMPA). This institution, a forerunner to Bach's institute in Stuttgart, advanced rapidly. The orders for testing rose from 525 in 1880 to 13,552 eight years later.[65] When the *Internationaler Verband für die Materialprüfungen der Technik* [International organization for testing materials] was founded in 1895, Tetmajer served as its first president.

Quite in line with what we described as the hybrid approach to mathematization, Tetmajer not only made a name for himself for accurate testing but also aimed to cast his results into mathematical expressions of general applicability. His best-known achievement is "Tetmajer's equation" (proposed in 1886) that describes the buckling of rods. At the time, engineers designed ever more daring constructions with new materials, in particular with a growing variety of iron and steel. These designs crucially required engineers to predict buckling—that is, secure the stability of a construction. Received wisdom held that Euler's equation covers buckling. Tetmajer pointed out that this is correct only for the elastic case, whereas rods under high pressure might buckle in a significantly different way. In fact, Tetmajer conducted extensive tests of building materials and synthesized his results into a law of buckling named after him. Much as in Bach's case, this law entails parameters that were introduced not for theoretical reasons but to make the law cohere to empirical data—that is, these parameters were adjusted to results from empirical testing in the laboratory.

A tragic accident illustrates the importance of buckling. On the splendid summer day of June 14, 1891, a crowd of more than five hundred people got on a train in Basel, Switzerland, heading for a festival at Delémont. Close to the town of Münchenstein, the train traversed a girder bridge over the river Birs when the construction failed and part of the train fell through the bridge into the river (see figure 3.7), killing seventy-three people and injuring more than one hundred.[66] Tetmajer was commissioned to examine the causes of this catastrophic failure.

The bridge had already experienced some modifications. It had been designed and built in 1875 by Gustave Eiffel's company that had mastered slender steel constructions.[67] After a flood in 1881 and later displacements of abutments that showed cracks, the bridge rested on (only) three instead of the original four piers, and several parts had been replaced and strengthened additionally. On the day of the accident, the train was so crowded that a second locomotive and extra cars had been added. Nonetheless, the bridge should have been safe—in theory.

Tetmajer found that something was wrong with the theory of statics. It was exactly that which he had discovered five years earlier in 1886: the inadequacy of Euler's equation in the inelastic case that was all too relevant for the bridge's stability. The girders of the bridge were under such heavy stress that the inelastic case set in when the train crossed. Use of his own

Figure 3.7
The second locomotive in the river. Courtesy: Wikimedia commons.

equation, Tetmajer found, could have predicted that the bridge was insufficiently strong.

Whereas the usefulness of Tetmajer's equation was widely acknowledged, its status as a "law" of buckling was debated controversially. An exchange in the *Schweizerische Bauzeitung* of 1895 between engineers G. Mantel, Felix Jasinski (France), and Friedrich Engesser (Germany) among others presents the point. Mantel discussed buckling beyond the limit of elasticity; Jasinski pointed out that Tetmajer deserved the credit; Engesser then came forward with his own formula that would put Tetmajer's equation on a rational basis—namely, it gave a simpler expression that was mathematically close to Tetmajer's; and in the limit, it recovered Euler's equation. Jansinski, however, replied that Engesser had made idealizing assumptions that are not justified empirically and lead to wrong values for buckling effects. In other words, Engesser qualified Tetmajer's hybrid approach as a heuristic phase before the theory would once again reach rational ground. Tetmajer disagreed. When he summarized his findings on buckling in a monograph on the laws of buckling (Tetmajer 1901), he insisted that mathematization should faithfully reproduce the experimental findings even if that foreclosed a coherent (rational) form. In other words, Tetmajer advocated a hybrid approach that combined rational and empirical elements but did not perceive a hybrid expression such as his law of buckling as a preliminary or heuristic stage on the road to fuller rationality. For him, the hybrid type of mathematization was the adequate approach because useful predictions depended on it—and because the crucial point was predictions rather than coherence to a wider theoretical edifice.

Steinmetz's Law

Charles Proteus Steinmetz (1865–1923) was a German-born mathematician and electrical engineer who emigrated to the United States. Edison and General Electric bought him (including his patents) out from a New York company, and Steinmetz became the leading engineer in Schenectady, New York, the major research and development site of General Electric. He was also known as the "Wizard of Schenectady," holding more than two hundred patents. His trademark was the use of mathematics to solve design problems in AC (alternating current) technology that was new at the time. The Charles Proteus Steinmetz Award is one of the highest honors the IEEE assigns.[68] Steinmetz gained his greatest prominence for two achievements: the theory of AC circuits and, in particular, the law of hysteresis, also called Steinmetz's law.

What is the matter with hysteresis? For engineers in the 1880s, ever more powerful AC systems had revealed that Maxwell's assumption of constant coefficients was wrong—that is, it ceased to produce adequate predictions for design. Hysteresis, meaning resistance to (de)magnetization, names the most important phenomenon in this context. Such resistance would lead to overheating and, therefore, energy loss. However, transformers depended on high efficiency, and it was crucial to design transformers that would avoid hysteresis as far as possible.[69] But if constant coefficients would not do, what else could replace them?

Practitioners relied mostly on tables that noted energy loss in some samples. From there, they tried to extrapolate by rules of thumb—with little success. Steinmetz started an attack strikingly similar to what occurred in Bach's case. He launched an extensive program of generating relevant data by inventing new high-throughput measurement instruments with a variety of materials, and he combined these with a sophisticated agenda to numerically integrate the evidence into a formula. Again, akin to Bach, Steinmetz introduced a new form with more parameters, one of them later called the "Steinmetz coefficient of hysteresis." After gathering data for more than two years, Steinmetz achieved a relationship that fit the empirical data with only a small error. In 1890, he published his equation in *Electrical World* as the "law of hysteresis."

On the one side, this law was a major breakthrough for design, and it was praised immediately by the engineering community. In 1892, the IEEE assessed that no paper of "more absorbing interest and practical utility has been presented to the [American] Institute [of Electrical Engineers] . . . We are sure that electrical engineers will feel a sense of relief in having finally gotten rid of another factor of uncertainty in the designing" (*Electrical Engineer* [Vol. 13, 1892, 87], cited in Kline [1992, 52]).

On the other side, the status of Steinmetz's findings was questioned. Was it justified to speak of the law of hysteresis? Physicists doubted whether the equation was sufficiently universal because it rested on data analysis rather than derivation. Ewing, a leading theoretician, found that the exponent varies between 1.475 and 1.9; and, therefore, the "law" should rather be called a useful design rule. Engineers such as Steinmetz were asking for accurate predictions in a technologically relevant range. Steinmetz finally admitted (as Bach did in the case of elasticity) that hysteresis is a complex phenomenon that his equation (setting the exponent to 1.6) described only closely, but not perfectly. He stopped calling it a law. Again, as in Bach's and Tetmajer's cases, the pivotal aspect was making adequate predictions (under certain conditions such as available materials) with mathematical means—that is, design equations. These equations contained rational (Maxwell) and empirical (parameterization of hysteresis) components. Only such a combination yielded predictions useful for design. Hence, hybridization was crucial. And faced with the choice of establishing continuity with existing theory or insisting on the practical usefulness of the predictions, Steinmetz (like Bach and Tetmajer) chose the latter option. This ties the hybrid mode of mathematization to prediction.

Heaviside's Operational Calculus

Our third and final case is again from electrical engineering. Oliver Heaviside (1850–1925) had an outsider career in England; he started to work as a telegraphist and electrician while teaching himself mathematics, physics, and electrical engineering.[70] Although James Clerk Maxwell is famous for the theory of electrodynamics, often referred to as "Maxwell's equations," it was Heaviside who used these four equations to express Maxwell's theory (Hunt 1991, 245–247), thereby attempting to provide Maxwell's theory with a more accessible and practically useful form.

In Heaviside's time, electricity became a major factor in industry and society, with Heaviside himself working on long-distance telegraphs. How Maxwell's theory would be relevant for electrical engineering was not at all clear. Could predictions about electrical systems be derived at all? Heaviside took on the task of offering a positive answer to this question. He used vector analysis and developed what he called "operational calculus"—a famously cumbersome mathematical technique for attacking "physical problems of technological importance" (Nahin 1988, 218). His calculus was good for obtaining predictions, but some of the symbolic abstracts were not defined in a mathematically rigorous way. On the contrary, he had to experiment with his calculus to see whether something meaningful would come out of it. In a way, Heaviside treated his calculus like a rule with some adjustable parameters. After adjusting parameters to observed cases, his calculus would be able to predict. The electrical engineer W. E. Sumpner (1928) described Heaviside's mindset in friendly words: "He was convinced about results as soon as he could verify them by severe experimental tests, and passed on without waiting to find formal proofs. He was a wanderer in the wilds and loved country far beyond railhead" (405).

As a result, Heaviside faced fierce opposition from mathematicians and physicists. Tait, for example, held that quaternions were the right notion, not the somewhat more tool-like vector analysis.[71] Although he became a fellow of the Royal Society, Heaviside struggled to gain recognition from the scientific establishment. His paper series (1892; 1893) "On operators in physical mathematics," which included years of experimentation with operators, ended abruptly after Part II because the Royal Society refused to publish more (normally, papers by fellows were not reviewed at all).

Heaviside defended a hybrid notion of mathematization. Because predictions of technological relevance were the goal, mathematization would be well advised to take in empirical information. "Mathematics is an experimental science, and definitions do not come first, but later on. They make themselves when the nature of the subject has developed itself. It would be absurd to lay down the law beforehand" (Nahin 1988, 222–223). Yavetz

(1995) provides a case in point when he shows that Heaviside's distortionless condition did *not* come out of a purely mathematical analysis. Empirical results showed the way—namely, "counterbalancing effects of inductance, capacitance, and resistance in a leaky transmission line" (213). This sort of counterbalance is well in line with the hybrid mode of mathematization discussed here.[72]

Despite their different subject matter, these three vignettes on Tetmajer, Steinmetz, and Heaviside reveal striking similarities with Bach's work when it comes to the hybrid mode of prediction. In the late nineteenth century, developing, using, and defending this hybrid mode was a key component when engineers argued for the autonomy of engineering knowledge. Establishing engineering science meant creating a hybrid culture of prediction.

This hybrid culture is not limited to the period studied here. For example, Vincenti (1990) impressively describes how, in the mid-twentieth century, engineers combined experimental measurements and mathematical formalization to design the shape of an airfoil. What Vincenti describes—in particular, the strategies of modifying and adjusting mathematical expressions—strongly resembles the cases considered here. The use of computers intensifies work with adjustable parameters in a remarkable way, even making it the linchpin of a new culture of prediction, as the following chapters will show.

Given that this standpoint was criticized mainly from the side of the rational mode, does that mean that the hybrid mode is (partly) irrational? Too much oriented toward prediction? We do not think so—for two reasons that both have to do with the coevolution of rationality and instrumentation. First, detailed studies of mathematical tools document that the perceived rationality is at odds with scientific practice. To pick out one exemplar of philosophical work, Wilson (2006) argues that the homogeneous classical picture of mathematical instruments dissolves on closer inspection into a facade of different patches (see also chapters 7 and 8). The second reason concerns the concept of rationality itself: when the hybrid culture of prediction evolved (interdependently on levels of knowledge, method, education, and institution), it also changed the concept of rationality, gearing it toward prediction.

II Toward an Epistemology of Iteration

4 Overlapping Modes in the Behavior of Molecules

This chapter takes the bull of prediction by the horns. One of the most formidable challenges is to predict properties of molecules by mathematical means, and one can rightly call this a candidate for the grail of prediction. Not only is the number of molecules, even in a small piece of matter, of a size at the edge of human imagination, but the otherwise successful strategy of idealizing simplification also reaches its limits. To compute how electrons interact in even a small number of molecules is still a task of nearly insurmountable computational complexity. In this area, the key to mathematization has been iterative strategies. Of course, such strategies predate electronic computers, but the vastly increased speed of computation that digital computers allow has opened up a new chapter in the coevolution of prediction, technology, and methodology.

We follow the trajectory of quantum chemistry that has been influenced strongly not only by mathematical techniques, especially iterative methods, but also by information technologies. The history of quantum chemistry covers all four modes of prediction—the rational and the empirical, and then the iterative–numerical and the exploratory–iterative—spanning pre-computer and computer times. Although there are not many notions that sound more technical than "quantum chemistry (QC)," the following text will use a minimum of technical language, and passages that refer to mathematical arguments can be skipped without losing the line of argument. However, the study of QC makes it necessary to also consider the epistemology of iteration.[1] To disentangle the concepts involved, we begin this chapter with a brief introduction to prediction and iteration (sections 4.1 and 4.2) as a teaser for the epistemology of iteration, before turning toward QC from section 4.3 onward.

4.1 A Pioneer of Prediction

In the year 1917, in the French region of Champagne near the front lines of World War I, an ambulance unit took a break inside a cold and wet rest billet. The driver, Lewis Fry Richardson (1881–1953), a sportive young man, sat on a heap of hay and immersed himself in a different world. In full concentration, his eyes behind small metal-rimmed glasses, he filled page after page over week after week with rows and columns of numbers following an elaborate numerical scheme. "It is a source of wonder that in such appalling nonhuman conditions he had the buoyancy of spirit to carry out one of the most remarkable and prodigious calculational feats ever accomplished" (Lynch 1993, 69).[2] Richardson was computing the first weather prediction based on quantitative physical laws. In fact, he was finishing the revision of his book manuscript *Weather Prediction by Numerical Process* (Richardson 1922) when he started to serve in the ambulance unit. In this manuscript, he had devised a new numerical way to tackle a mathematical model of atmospheric dynamics. The large majority of his meteorological colleagues had deemed such an effort impractical. For them, theoretical meteorology was a topic in which one dealt with theoretical models and physical laws, whereas predicting the weather was a different topic that had to be approached by entirely different means.

The work of the eminent meteorologist Vilhelm Bjerknes illustrates this split between theory and prediction. On the one hand, he is known for his theoretical work putting together a system of fundamental equations for global atmospheric dynamics (Bjerknes 1904). These equations are all well justified from the perspective of physics (motion and conservation laws), but they form a system that is impossible to solve analytically. On the other hand, Bjerknes did not even try to work with this system of equations for purposes of prediction. Instead, he devised a way to analyze the development of weather fronts graphically on the basis of empirical data.[3]

Early on in his scientific career, Richardson had decided to work on the possibility of turning mathematical equations into predictive devices. Though perhaps idealistic, Richardson was fully aware that success depended on a difficult combination of measurements and mathematatization. He turned away from the rational–analytical perspective on mathematics and designed a computational scheme based on iteration. What he was after was a groundbreaking epistemological turn toward the iteration necessary when working with a discrete, noncontinuous model. In it, the atmosphere would evolve stepwise on a space–time grid, and the calculus would be replaced by an iterative finite difference algorithm.[4] Crucially, such a procedure would get rid of the complicated integrals that described the evolution of the system. They would be replaced by simple computational steps that had to be iterated repeatedly at each of the grid points.

This iterative mathematical procedure, however, had to be based on developments in theory as well as in empirical knowledge. In the theory of turbulence, especially the transport of eddies, Richardson had to find reasonable and yet unknown means of approximate treatment. Such approximation schemes could not be justified strictly from theory, and their results had to hold their ground against empirical measurements. Hence, finding reasonable paths of approximation required accurate empirical knowledge about eddy transport. Such knowledge, however, did not yet exist and had to be based on new measurement techniques for the higher atmosphere in order to obtain values for the initial conditions.

Hence, by 1917, there were doubts about whether accurate weather prediction was possible at all, and Richardson accepted that. However, he could erase those doubts by actually making predictions. He decided to add a chapter with a prediction obtained by the methods and numerical schemes explained in the main parts of his book. He showed his characteristic hardiness in carrying out the calculations under the conditions just described. The prediction used one grid cell of several hundred square kilometers in Germany, drawing on known data (including very low-quality estimations of conditions in the higher atmosphere) from May 10, 1910. Richardson started with the wind and air pressure conditions at 7 a.m. and calculated the conditions six hours later. The result was catastrophically wrong. Somewhat surprisingly, Richardson included the example anyway. This decision is telling for two reasons: first, it was motivated by the sheer difficulty of obtaining any predicted values at all because the computation took so long that Richardson did not have the option of starting with a different set of conditions. Second, Richardson tolerated the wrong values because he could argue that he proved the principal possibility of weather prediction.[5] Richardson's approach via iteration and numerical approximation was totally unusual at the time.[6] Both the culture of mathematics and that of meteorology were biased against it. Mathematicians were not willing to include what they considered to be low-quality "approximate mathematics" into their disciplinary canon because they objected to trading theoretically demanding integrals for iterations whose difficulty arose from their great number rather than their sophistication. Thus, Richardson's finite difference approach was not accepted by the mathematical faculty at Cambridge (Hunt 1998, xvi). Meteorologists relied on entirely different empirical and graphical approaches to prediction. Consequently, Richardson's book that appeared in 1922 had a more or less negligible impact and was criticized for mixing up theoretical results about turbulence with computational schemes.[7]

Richardson fully realized that for his computational scheme to be practical, it had to satisfy two conditions: (1) computations had to be completed quickly enough for the prediction to still be about the future; and (2) observations and computations together had to be socially affordable. He reasons in the preface: "Perhaps some day in the dim future it will be possible to advance the computations faster than the weather advances and at a cost less than the saving to mankind due to the information gained. But that is a dream" (Richardson 1922, xi).

Even if considered a dream, this vision stimulated Richardson (1922) to speculate on the organization of computing:

> Imagine a large hall like a theatre, except that the circles and galleries go right round through the space usually occupied by the stage. The walls of this chamber are painted to form a map of the globe. The ceiling represents the north polar regions, England is in the gallery . . . A myriad computers are at work upon the weather of the part of the map where each sits, but each computer attends only to one equation or part of an equation. [On a tall pillar in the middle, in the center of the spherical room] . . . sits a man in charge of the whole theatre; he is surrounded by several assistants and messengers. One of his duties is to maintain a uniform speed of progress in all parts of the globe. In this respect he is like the conductor of an orchestra in which the instruments are slide rules and calculating machines. (219)

Nowadays, Richardson's book is considered to be among the most prominent monographs on weather prediction and an inspiration for computational scientists. After electronic computers had changed the technology of computation, Richardson's vision looked much more realistic. When George Platzman (1967)[8] re-reviewed Richardson's book in 1967, he expected even more than Richardson had dreamed of:

> It seems likely that the whole of physics and chemistry, with all their ramifications, will in time be reduced to mathematics, enabling the entire future course of

material events to be predicted if only our powers of observation and calculation were equal to the task. (520)

In other words, prediction will be mathematized on the grandest scale of science if only two conditions fall into place: big data and computational power. It is remarkable that computational power is identified as a bottleneck exactly after the computer had magnified this power enormously. Platzman already lived in a new culture of prediction whose use and needs of resources had changed.[9] When the digital electronic computer became part of a culture of prediction, the concept of iteration acquired a central role. In the following, we shall set out the epistemology of iteration, investigate its close relationship to the digital computer, and scrutinize the ways in which "small" computers induced an exploratory element to the epistemology of iteration. As we shall see, this epistemology will also cast doubt on the claims about reduction that Platzman (and many others) took for granted.

4.2 A Rose Is a Rose Is a Rose

Iteration is a concept present in nearly all domains of human activity. Solving a problem by trial and error is a strategy based on repeated attempts. Exercises in music such as playing the violin often consist in doing the same thing over and over—which later might lead to a new level of performance and understanding. Iteration—simply put: doing the same thing over and over—has surprisingly far-reaching epistemological ramifications.[10] This chapter concentrates on iteration in a mathematical (and computer) context in which formal operations are repeated. A simple example is the multiplication $n \times 7$. For $n = 3$, readers know the answer since childhood. For a bigger n, say $n = 268$, the answer is less obvious, but calculation strategies from elementary school will lead to the answer in a straightforward way. One way to do this is to follow an iterative algorithm: start with zero, add 7, and do it again until you have done it n times. This is an awkward algorithm for a human, but a tailor-made exercise for a digital computer. How can one characterize the epistemic significance of iteration?

Here are two observations from the multiplication example: first, what appears trivial on the conceptual side is all but trivial on the practical side because the abilities of human operators are limited. Consequently, strategies are attractive that reduce the load of iteration. Second, iteration feeds into automation by computing machines. Accordingly, numerical strategies pivot in the opposite direction and try to transform conceptually complicated operations into a series of simple operations—that are then carried out by a machine.

Differential and integral calculus can serve as a second, more advanced, illustration. Leibniz, one of the main inventors of calculus, was concerned with determining properties of mathematical curves (functions). Normally, this required tedious and complicated geometrical constructions. He aimed to devise some rules that would allow him to circumvent these constructions—much like the way algebra, by calculating with letters, is able to shortcut arithmetical calculation with numbers. His calculus became a major vehicle for mathematization in physics. Like all instruments, it has its limitations. Often, dynamic systems can be described mathematically by a system of partial differential equations that is itself intractable. Calculus has paved a way for mathematization that (much later) leads to dead ends. This brings into play the second strategy, the one we already encountered with Richardson. He replaced differential equations with difference equations—that is, he split up intractable integrals into a great many additions. The second, iteration-friendly strategy was a dead end for classical analytical methods but turned into a broad avenue for numerical operations—if only iterations could be executed in sufficient numbers—which is a big if.

This strategy reached full bloom with the digital electronic computer, but it is much older. Early examples are procedures for approximation such as the Newton–Raphson method.[11] Their ingenuity lies in the balance they strike between replacing intractable problems by iterations of simpler steps and keeping the number of these steps manageable. The epistemology of iteration asks what happened when the iterative strategy was pushed further. Human computers working with pencil and paper can navigate a relatively narrow range of iterations, but things look different when algorithms are transferred to a nonhuman computer—that is, to a machine.[12] Charles Babbage (1791–1871) was convinced that his *Analytical Engine*, a versatile mechanical calculator that could be programmed, would open up new territory for mathematization.[13] Whereas Babbage's machine was never finished, much less versatile semiautomated mechanical desktop calculators were widely used in the second half of the nineteenth century. They allowed the cumulation of additions in large numbers.[14]

Notwithstanding this history of iterative strategies, it was the electronic digital computer that catapulted iteration into the center of mathematization.

Through technological innovations from vacuum tubes to transistors to ever faster integrated circuits, the speed of computations increased vastly, making iterative strategies feasible that were not even worth thinking about before. This had a major impact on mathematical modeling—and, the other way round, computing technology saw a constant stream of innovation. Technology, mathematization, and social organization coevolved and formed an iterative culture of prediction.

4.3 Quantum Chemistry: Competing Pathways

From here onward, this chapter follows the history of quantum chemistry (QC)—that is, the endeavors to computationally predict chemical properties of molecules.[15] Tracing this subject matter leads us through the study of different cultures of prediction. In principle, prediction faces daunting difficulties because chemical properties depend on the complex interactions of electrons. From our perspective, however, this is fortunate: in the attempt to access an almost inaccessible territory, mathematization, modeling, and technology coevolved in a very dynamic way.

QC spans all four different cultures of prediction addressed in this book. It starts prior to the computer with rational and empirical approaches competing with each other. We shall analyze two crucial turns in the trajectory of QC. Both are connected to new conceptions of computational modeling that developed in line with the availability of new computing technology. The first turn began in the 1950s when electronic digital computers became available, and it was completed around 1970. By then, QC was firmly rooted in an iterative–numerical culture of prediction,[16] often referred to as an "ab initio" approach. During the 1990s, QC underwent a second turn that is tied to the easy availability of small, networked computers. This second turn again led to the establishment of a new culture of prediction that we call exploratory–iterative.[17]

However, computational technology and infrastructure did not simply determine the path of QC; rather there existed—and there still exist—complementary conceptions of mathematical modeling that are in flux and closely related to available instrumentation. The computer has exerted a fundamental influence on the development of QC not just by changing computation in a technical sense but rather by leading to a rearticulation of the practices of the QC community. Rather than determining the development, computing technology played more the role of a boundary condition that informed the ongoing process of mathematization.

The birth of QC started in 1926 when Erwin Schrödinger (1887–1961) published his wave equation describing the energy of a quantum mechanical system. At the time, it was neither the only nor the first formulation of the new quantum mechanics, but Schrödinger's formulation as a wave equation used the idiom of calculus and thus resonated with the mathematical practices of chemists. It was not clear whether the potentially new field of QC, called "chemical physics" at the time, would lean more toward chemistry or physics.[18] There were two complementary views of the intersection of physics and chemistry: the first can be called "principled theory" and foregrounded the physics side, whereas the second is often denoted as "semiempirical" and brought in substantial experimental traditions from chemistry. Both flourished from the start. Let us start by turning to the first view.

The Rational Standpoint

The principled theory view holds that the Schrödinger equation contains all the information about the electronic structure of molecules that determines a system's chemical properties such as the bond energy levels. Consequently, the equation slips into the role of a fundamental mathematical law that governs significant parts of chemistry. This is the foundational step toward a mathematization of chemistry, a mathematization that holds as much promise as rational mechanics did for physics.

However, the rational viewpoint had to face considerable difficulties due to the mathematical form of the Schrödinger equation. This equation expresses a system's energy via a wave function $\Psi (1, 2, \ldots, N)$ that has as variables all (N) electrons of an atom, molecule, or bunch of molecules. The electrons interact, and hence Ψ has $3N$ degrees of freedom (i.e., three dimensions of space, leaving spin aside)—a number of discouraging cardinality. Even if one restricts the focus to only a few electrons and their interaction, solving Ψ is extremely difficult and computationally demanding—indeed, practically impossible beyond the simplest cases. Therefore, it was an open question whether the principled possibility of prediction would translate into a mathematical approach that could, in fact, predict relevant properties.

A positive answer to that question came fairly quickly. In 1927, a joint paper by the German physicists Walter Heitler and Fritz London argued that some chemical bonds could be understood as a quantum phenomenon

(Heitler and London 1927). The argument was of a mathematical nature: the Schrödinger equation implied that two hydrogen atoms reduced the total energy when they formed a bond. Although the energy that Heitler and London calculated from the equation was not very close to the energy known from experimentation, they took their result as a great success. Chemical attraction, the computation showed, rested on electron exchange—that is, on a quantum-mechanical effect. For Heitler and London, this insight was far more important than the accuracy of the quantitative value (cf. Nye 1993). This first mathematical result was immediately recognized as a proof of principle: mathematical prediction from the Schrödinger equation was possible.

Quickly, a small group of quantum theorists, mainly physicists, became convinced that pursuing this path further would lead to a new QC. The next steps were to tackle (slightly) more complicated cases and to produce more accurate values. Egil Hylleraas, a Norwegian physicist working as a postdoc with Max Born, devised an approximation of the two-electron system of helium and calculated values for ionization energy that were very similar to those measured experimentally. This was taken as numerical proof that quantum mechanics does indeed govern chemical properties, and that mathematical prediction could be accurate enough to be chemically relevant (Park 2009).

One should note, however, that these results were retrodictions—that is, mathematical derivations aimed at matching results already known from experiments. Hylleraas's goal was to find mathematically plausible approximations that fit the values known from experiments (what Park [2009, 34] calls the "practice of theory"). Thus, although the rational approach saw QC as a genuinely theoretical endeavor, when it came to concrete values and the question of accuracy, the supremacy of experimental results was not challenged.

The aspiration of the rational viewpoint is expressed succinctly in the physicist Paul Dirac's (1929) notorious quote that is cited in virtually every portrayal of QC:

> The underlying physical laws necessary for the mathematical theory of . . . the whole of chemistry are thus completely known and the difficulty is only that the exact application of these laws leads to equations much too complicated to be soluble. (714)

This quote has been received in interestingly different ways: on the one side, many quantum chemists today focus on the claim of nonfeasibility and find it overly pessimistic. New methods and strategies of computational modeling have been able to yield increasingly sophisticated approximations that make it easier to tackle these equations numerically. On the other side, many historians and philosophers have concentrated on the implied claim about the reduction of chemistry to quantum physics. Eric Scerri (1994), for instance, finds the Dirac quote to be too optimistic because it overstates the implications of "mathematical theory."[19]

Before long, the principled theory program ran into computation troubles that derailed mathematization for decades. In 1933, James and Coolidge reached a veritable impasse: they used trial functions—that is, building blocks for approximations—and were willing to add as many terms as were necessary to obtain an accurate fit. Eventually, while the quantitative results indeed looked satisfactory, it also became clear that the computations were forbidding—it took them about one year of intense work, including the use of mechanical computing devices, to finish them (Park 2009; Schaefer 1986). It turned out that pairs of electrons constitute very special and relatively easy cases, whereas existing numerical strategies would break down in other cases. Methods had focused on the outer electron shell for reasons of simplicity, but it appeared that by taking the inner shells into account, the approximations would actually become worse—the so-called "nightmare of inner shells" (Park 2009, 48).

Thus, the principled approach became stuck. The derivation from the fundamental laws—that is, from the Schrödinger equation—demanded effective approximation strategies, and there seemed to be no way to devise feasible procedures. Numerical strategies have to respect what amount of iteration is doable in a reasonable amount of time. For instance, a trial function includes terms to be fitted, and one can use many terms to reach a good approximation (as James and Coolidge were trying to do). But this procedure normally requires iterating the approximation for each term (rather than deriving the terms analytically). Hence, a good quantitative fit depends on the speed and ease of performing iterations. However, the necessary iterations were not manageable with the calculational means available to the researchers at that time: these problems seemed to be insurmountable. The mathematician and chemist Charles Coulson recalled in retrospect that after the early 1930s, the development of wave mechanics came to a full stop and "despondency set in" (Coulson, cited according to Nye [1993, 239]).

The Empirical Standpoint

A second view already complemented the first one early on. Right from the start, its proponents accepted that experimental approaches would be valuable resources. When computational procedures encountered components (integrals with physical meaning) too complicated to compute, one could first plug in values that had previously been determined experimentally and then continue with the procedure. In this way, empirical results helped to overcome—or rather circumvent—computational difficulties. In the context of QC, this approach is usually called semiempirical, and it did not encounter the impasse of the rational approach. "Devising semiempirical approximate methods became, therefore, a constitutive feature of quantum chemistry, at least in its formative years" (Simões 2003, 394).

Three persons are exemplary for this second strand: the chemist Linus Pauling, the chemist and physicist Robert Mulliken, and the physicist John Slater. All three were born and educated in the United States, and, despite differences in their disciplinary affiliations, they followed a more pragmatic approach than the adherents of the first strand.[20] All three worked in the 1920s as postdoctoral scholars with quantum theorists in Europe. Pauling visited Sommerfeld in Munich, Mulliken worked for Hund in Born's laboratory in Göttingen, and Slater visited Bohr in Copenhagen.

The semiempirical approach was oriented toward prediction and conceived of computational modeling as a pragmatic combination of theory and experimental results. The foremost characteristic of all the semiempirical approaches is that experimentally measured values are imported into calculations in order to make predictions. One might notice that both theory and experiment were crucial parts of semiempirical models, and that this seriously undermined claims of reduction to theory as expressed by Dirac.

A typical instance is Slater's proposal to approximate a molecular orbital. There were two competing approaches: Pauling promoted the valence bond approach that worked with localized individual electrons that built bonds to form molecules. Mulliken advocated the molecular orbital approach that assumed a mixture of uncorrelated electrons that were shared by one molecule. However, Slater proposed a nonlocalized orbital that belonged to a whole molecule by a linear combination of atomic orbitals. In mathematical terms, modeling the combination in linear terms offered maximum tractability and adjustability. In this way, Slater made the two complementary approaches computationally compatible—in the service of obtaining (i.e., predicting) accurate numerical values.

Often, the semiempirical approach is seen as an alternative or even a contradiction to the established culture of theoretical-mathematical physics. Coulson talks about the "Pauling era" to denote the turn in thinking when the pioneers of QC started to escape from the "thought forms of the physicist" (Coulson [1970, 259], cited according to Simões [2003, 396]). However, mathematical modeling also played an essential role in Pauling's approach. One illustration is the notion of "resonance" that did not aim to represent some factual phenomenon but served to achieve better predictions. Coulson diagnosed that "resonance is . . . a method of calculation; but it has no physical reality" (Coulson [1947, 47], cited according to Simões [2003, 398–399]). Pauling, who viewed chemistry as a human-made endeavor anyway, did not find the lack of a physical reality problematic—"unnatural" was not a criticism he cared much about. Mathematical modeling inevitably adds an element of artificiality to any system. Models are, in important respects, *dissimilar* to the objects and phenomena they claim to represent, much like the ways a map represents a territory in specific ways for specific purposes.[21]

The first, rational approach has much in common with what we called the rational culture of prediction (cf. chapter 2). In it, mathematization unfolds the implications of fundamental laws. The semiempirical approach adds flexibility to the process of mathematization, mixing representational and performance-oriented components much like we found in the empirical culture of prediction (chapter 2). Furthermore, a hybrid approach was able to produce predictions when the rational approach stalled, similar to what happened in late nineteenth-century mechanical engineering (chapter 3). This was only the beginning of QC's history. The rational approach revived under new circumstances, when iterative algorithms and automatization of computing set mathematization on a new track. The iterative–numerical culture of prediction flourished with the digital computer but has older roots that are worth inspecting.

4.4 Interlude: A Mechanico-Numerical Program

A central requirement of computational modeling is tractability. Computationally tractable models produce desired numerical values in a reasonable time given the computational means at hand. This imposes a particular condition on mathematical modeling because such models do not work on the basis of general mathematical description but call for concrete algorithms. For instance, a mathematical model might indicate that a certain value is defined uniquely, whereas a *computational* model also gives a procedure for determining this value in practice. Furthermore, tractability is not an absolute concept but depends on the instrumentation available, and algorithms tractable with one calculational tool may be intractable with another.

The mathematician Douglas R. Hartree (1897–1958) clearly acknowledged the importance of instrumentation for mathematical modeling. He combined great mathematical skills with a passion for tinkering and for the automation of calculation. For instance, he built a working copy of Vannevar Bush's differential analyzer out of Meccano parts that he filched from his children's toybox. Hartree was an early proponent of digital computing machines. As early as 1947, he articulated a vision of computing technology and mathematical methods as "twins" (Hartree 1984), and he pioneered the iterative–numerical mode of prediction.

In QC, Hartree is best known for the Hartree–Fock method going back to 1927 when he devised an approximation strategy for wave functions (and energies) of atoms. This strategy presented a bold move toward iteration at the expense of mathematically solving the Schrödinger equation. Hartree conceived the challenge in a new way, resolutely oriented toward finding an adequate prediction rather than an analytical solution. He prioritized computation (instead of mathematical solution), devised a class of numerical procedures that could be iterated semiautomatically (with the help of some mechanical devices), and then specified a model for which one of these procedures would work.[22]

Here is a brief outline of how Hartree used iteration to work around computational problems. The Schrödinger equation is so mathematically complex because each electron interacts with the others so that a solution cannot proceed by splitting up the problem and calculating the subproblems independently one after the other. Hartree's approach starts by calculating the value of the potential of one electron, thereby assuming an ad hoc initial value and counterfactually assuming all other potentials as fixed—that is, as a given field. In the next step, the first electron assumes the value calculated in the first step, then the second electron is regarded as variable and all others as fixed. Then the procedure is iterated. The first series terminates when each electron has played the role of the variable. Then, the next series of iterations starts by taking the results of the first round as initial conditions. At the end of the second series of iterations, the values of all electrons are readapted. If they differ from the first series, the whole procedure starts over a third time, and so on until the values no longer change between two series of iterations. In short, each step ignores part of the electron interactions to find values that are mutually consistent, hoping that errors then cancel each other out. This procedure is known as the self-consistent field (SCF) approach.[23]

One can criticize Hartree's approach as unprincipled and artificial, oriented merely toward numerical and mechanical feasibility. For Hartree, however, computational tractability outweighed the missing theoretical justification of the model. The SCF procedure partly ignores the interdependence of electrons—that is, the main obstacle in terms of computational complexity—whereby the procedure inevitably "mistreats" mutual interdependence. It trades predictive capacity against theoretical validity. Hartree's approach became widely accepted when experience showed that the predictions obtained by SCF—that is, Hartree–Fock—were good enough. Even today, Hartree–Fock methods are still in common use.

The lesson Hartree's SCF example teaches is as follows: if computational modeling aims to produce accurate enough predictions, modeling might develop according to its own demands and rationale—mediating between theory, experimental data, and computing technology but not determined by them.[24] Hartree's method took numerical feasibility as a guiding criterion for co-constructing the model and instrument. His major contribution, we would like to argue, is not mastering the challenge of numerical feasibility but approaching the model and the instrument as one project.

Hartree was not only an expert in computing technology in general but was also a very early proponent of digital machines in particular. He conducted pioneering work with the Electronic Numerical Integrator and Computer (ENIAC) in Pennsylvania and applied that experience to the development of the Electronic Delay Storage Automatic Calculator (EDSAC) computer at Cambridge University. Early on, Hartree (1949, 1958) realized from his experiences with the ENIAC and its general-purpose programmability that digital computing would not merely make computation faster but would open up a new path for mathematization. Mathematical modeling and computational technology would coevolve: "It is necessary not only to design machines for the mathematics, but also to develop a new mathematics for the machines" (Hartree 1949, 115).[25]

4.5 Organizing Computation in the Iterative–Numerical Mode

The development of the iterative–numerical mode of prediction preceded the digital computer; and, moreover, the eventual establishment of a culture of prediction did not happen in one fell swoop. Rather, it was the outcome of an ongoing dynamics: how computation was organized institutionally and technologically.

The advent of the digital computer did not change everything completely, nor did it result in a new start for computational chemistry. Rather, digital electronic computers transformed already existing computational strategies such as the one heralded by Hartree: ones that worked with a moderate amount of iteration and that were designed for a mechanical integration technology. With the digital computer, algorithms that required many more iterative steps became feasible. Since the early 1950s, computational modeling had been seen increasingly as a topic that demanded special attention from quantum chemists. A small community formed that was located mainly in the United States and the United Kingdom. It included the arguably leading group of Robert S. Mulliken and Clemens C. J. Roothaan at Chicago and the group of S. Francis Boys at Cambridge.

This community shared the belief that the conditions, possibilities, and limitations of computation were main factors for development. In other words, QC should not be determined by its theoretical object alone; rather, it should codevelop with computational instrumentation. The automation of computing was no longer viewed as simply an extension of human mathematical skills but rather as a different approach to mathematization altogether. In our example, this was privileging iterative–numerical procedures over analytic solutions to partial differential equations.[26] This new conviction was firmly rooted even before the digital computer became easily accessible for chemists. The first achievements of computational modeling in chemistry were not so much predictions of chemical interest—in this respect, experiments still held the leading role, and computational accounts still aimed at retroacting known results—but the development and investigation of systematic foundations of computational modeling (cf. Schaefer 1988).

Computational aspects influenced mathematical modeling quite fundamentally. A telling example is Boys's (1950) introduction of a special class of basis functions for approximating orbitals called Gaussian functions because they belong to the same family as the Gaussian distribution (i.e., the "bell curve") function. The agenda of Boys's group was to treat chemically interesting (i.e., bigger) molecules. Hence, computational virtues of models became a focus of research. It is important to recognize that the concept of "computational virtue" is dynamic in two different ways: first, the virtue of feasibility—that is, avoiding the computationally intractable—depends on the available computing technologies. Second, not unlike other virtues, it has a social character: at stake is the performance of models in those particular fields or on those particular problems that the community deems interesting—and that itself is dynamic. Boys was intrigued by the computational properties of Gaussian functions that made it possible to tackle multidimensional integrals with relative ease. He established these functions as a means to approximate the computationally much less tractable exponential or Slater-type orbitals, although it was known that such treatment introduced extra errors because of the different form of Gaussians. However, Boys preferred computational tractability over quantum-theoretical justification.[27] Soon it became clear that the use of Gaussian basis functions rendered tractable a range of molecules that formerly had been out of reach. This led to their widespread adoption along with the possibility that the accompanying approximation errors could be addressed via correction factors.[28]

The community's computational agenda was documented at a 1951 conference on Shelter Island, New York. Mulliken gathered most of the leading people working on QC at the time for a workshop with National Academy of Science support. Robert Parr and Bryce Crawford produced the official report of the conference: "National Academy of Sciences Conference on Quantum Mechanical Methods in Valence Theory" (1952).[29] The rationale for the conference was the shared opinion that computational problems were a major obstacle in resolving inadequacies in valence theory. Parr (1990), who attended as a young researcher, recalled later that the members of this community were so committed to tackling computational problems that they discussed "calculation as a way of life to be adopted by us" (327).

The Shelter Island group was to make a joint and organized effort by systematically assigning and distributing salient computational tasks, thus trying to combine the relevant with the feasible. The envisioned strategy comprised two steps: First, a working group determined which of the mathematical integrals that created computational problems were also of general significance—that is, occurred in a range of models. Such integrals could then play the role of building blocks for models in QC. The second step consisted in actually evaluating these integrals and producing a collection of numerical tables to be a shared asset of the QC community. "An informal Integrals Committee was established, centered at the University of Chicago, to collect and dispense integral information" (Parr and Crawford 1952, 552). We see here the intention to standardize computational models and to organize a distributive mode for computational modeling.

4.6 Infrastructure and Bottlenecks in the Iterative–Numerical Culture of Prediction

Progress in computational modeling gave the rational approach a boost, though a limited one. The attempts to calculate or numerically approximate solutions to the Schrödinger equation—what we called the rational approach—had come to a halt in the early 1930s. The semiempirical approach dominated from the 1930s onward, not least because it could profit directly from advances in measurement and experimentation. However, on the basis of the advances in computational modeling made in the early 1950s, the prospects for a relaunch of the rational approach began to look more hopeful. One of the first new calculations was reportedly finished by Charles W. Scherr in 1955 who calculated the energy of the N_2 molecule. It took him and two assistants a year with a desk calculator (see Park [2003] for more details). Whereas Scherr profited from advances in computational modeling, his approach still consumed so much time that it remained largely impractical. Although computational strategies had turned complicated mathematics into iterations, the available technology was still too slow.[30]

Quantum chemists were aware that, in principle, the use of digital computers held promise for an iterative–numerical mode of prediction. In practice, however, using the technology created more of a bottleneck. In the mid-1950s, only a small number of digital computers existed. They were extremely expensive and were owned mainly by government agencies, meaning that access was difficult for such scientists as quantum chemists.

Additionally, these machines were difficult to use in lieu of powerful compilers and other software. Nevertheless, quantum chemists tried to get access to these machines early on.

Bernard Ransil's pioneering work illustrates both the promise and the limitations of digital computing. Ransil was computing wave functions of the H_3 radical with the SEAC, the Standards Eastern Automatic Computer, one of the first electronic computers housed at the National Bureau of Standards. The SEAC was the first fully functional stored program computer in the United States. In 1955, with a newly minted PhD in physics, Ransil moved to the University of Chicago to join Mulliken and Roothaan's QC group. There he worked on nitrogen, specifically on the design and construction of the "first computer program to generate diatomic wave functions in minimal orbital LCAO-MO-SCF approximation" (Bolcer and Hermann 1994, 8).[31] Ransil wrote this program in machine language for a UNIVAC 1103 computer located at Wright Field Air Force Base in Dayton, Ohio. Mulliken and Roothaan, eager to gain access to digital computing facilities, had contracted computing time from the military. The infrastructure was not exactly inviting. Ransil had to prepare the set of commands, travel from Chicago to Ohio with a stack of prepared punch cards, and work overnight with the UNIVAC (see Mulliken 1989). The modification of the program was extremely tedious by today's standards due to the working conditions—working in Chicago, debugging in Ohio—and also due to the technical conditions of computation. In particular, machine language programs did not offer the relative convenience of an optimizing compiler such as FORTRAN (first available in 1957 from IBM), so programming equations was much more tedious (requiring many, perhaps twenty times, more statements than FORTRAN would); and, furthermore, any modification of the model would regularly require substantial new programming. Nevertheless, the program ran eventually, and Ransil obtained the desired value.

Mulliken and Roothaan, who had looked eagerly over Ransil's shoulder, considered this a breakthrough and shouted the result from the rooftops in "Broken Bottlenecks and the Future of Molecular Quantum Mechanics" (1959). The bottleneck to which they referred was the complete automation of a computational procedure. They reported Ransil's machine program that calculated important aspects of wave functions of diatomic molecules. "The importance of such a machine program is illustrated by the fact that the entire set of calculations on the N_2 molecule which took Scherr (with the help of two assistants) about a year, can now be repeated in 35 min" (Mulliken and Roothaan 1959, 396). Of course, the speed of computation played a major role. Nonetheless, taking advantage of this speed presupposed conceptual as well as organizational conditions—namely, accessibility and usability of the technology and the right kind of computational models geared toward iteration. We see here an early exemplar of the emerging iterative–numerical culture of prediction.

Our example involves two types of iteration: first, the iterations of arithmetical or logical operations performed by the computer—that is, algorithmic iteration. The second type of iteration we call interactive. It involves a feedback loop between computer, model performance, and researcher. Interactive iteration was difficult in Ransil's case because of the way the computing instrument was accessed. Interactive iteration demanded time-consuming, repeated trials to make the software run successfully, and this also required trips between Chicago and Dayton plus negotiated access to the computer at Wright Field. Exploring modifications of the program was practically infeasible—too costly in terms of time and workload. Further exploratory work would have demanded an infrastructure in which going back and forth between running and modifying the model was a practical option. Therefore, although the bottleneck for algorithmic iteration was broken by work like Ransil's, interactive iteration remained largely blocked.

An additional illustration of how computational infrastructure is related to, or rather prohibits, interactive iteration comes from West Germany. Digital computing entered science somewhat later in Germany than in the United States. The *Deutsches Rechenzentrum* (DRZ), the German Center for Computation located in Darmstadt, bought a "central" digital computer only in 1961 and made it available to German universities. The first machine was an IBM 704, replaced by an IBM 7090 in 1963. German quantum chemist Sigrid Peyerimhoff recalls that back in the mid-1960s, one had to travel to Darmstadt with a box of punch cards or prepared magnetic tape. There, one handed over the cards to the staff at the DRZ who ran the program and returned the output. Of course, modifications had to be made frequently to debug the program. Thus, to avoid frequent traveling, successful researchers needed to know a DRZ staff person who could oversee the process and was willing to take phone instructions. One could also send the punch cards by mail, but that would further slow down the whole process—debugging always required multiple attempts. Output was always sent back by post.

Such conditions allowed for only rudimentary interactive iterations. Peyerimhoff (2002) experienced how limiting the infrastructural bottleneck actually was during a stay in the United States (speaking about 1963):

> In Chicago, I realized for the first time how important it was to have access to a reasonably sized computer (IBM 7090) on campus, even if runs could be performed only during the night. And furthermore, that a turnaround time of a day or two for computer jobs made all the difference compared to the German situation using a Z23 [Zuse type 23] or sending programs and outputs back and forth to the DRZ in Darmstadt by regular mail. (271)

Whereas interactive iteration was challenging from a practical point of view, it also formed an essential part of computational modeling. Of course, no programmer can escape debugging because it is an iterative process. Even more importantly, interactive iteration is at the core of parameter variation when some model has a number of parameters whose values are deliberately left open and assigned later depending on the performance of the model.[32]

The restricted accessibility of computing machines was widely seen by practitioners as presenting a major obstacle for QC because it prevented parameter variation as a working practice. In a famous passage from Mulliken's Nobel lecture in 1966, he articulated the promise that computational modeling held for chemistry, but also identified one major caveat: "There is only one obstacle, namely that someone must pay for the computing time." Then Mulliken called for government support. Thus, the epistemology of iteration is linked to technology and infrastructure: whereas the bottleneck of mechanizing (algorithmic) iterations was broken, cost and ease of access emerged as new bottlenecks that prohibited interactive iterations.

Indeed, in the United States, there was one group that strongly advocated for the establishment of a computing center for chemistry: the National Resource for Computation in Chemistry (NRCC).[33] In 1964, the Lawrence Berkeley Laboratory received one of the first supercomputers, the CDC 6600 designed by Seymour Cray. From 1971 onward, it was made available to other users who were not members of the lab—mainly users working in high-energy physics but also some outside users such as chemists. Practically, this was similar to the West German DRZ—getting onsite instruction, mailing magnetic tapes, sending instructions via terminal, and receiving printouts via paper mail. This was the reason for calling for their own center: quantum chemists wanted more access, not leftover computing time from nuclear research.

But the exact structure and form of an NRCC was not determined. Should the NRCC follow the lead of physics and aim toward supercomputing? Right from the beginning, the proposal was not favored unanimously among chemists. In particular, Michael Dewar opposed these plans—computing time on a supercomputer facility was very expensive and "with so much pressure to justify results, Dewar feared that extensive calculations of an exploratory nature would be too greatly inhibited" (according to NAS [1971, 45]). Moreover, Dewar worried about a potential bias against semiempirical work that had proved so useful for QC. The mere presence of a supercomputer would potentially direct modelers toward formulating complex models that, in turn, would require such a computer. Dewar advocated a network of smaller and cheaper computers instead. This was an approach that was ahead of his time and would not materialize for another two-plus decades.

The controversy about the exact nature of an NRCC can be interpreted in light of the epistemology of iteration: on the one side, if algorithmic iteration were to be accelerated so greatly, computational models of a new size could become tractable. To achieve this seemed to require supercomputer infrastructure. On the other side, critics such as Dewar stressed the importance of interactive iteration for exploratory computational modeling. This demanded cheap and easy access. Hence, the two types of iteration demanded different infrastructures and were thus at cross purposes.

In 1977 the NRCC was finally established as a division of the Lawrence Berkeley Laboratory. The National Science Foundation and Department of Energy, the joint sponsors of an annual budget of $1.75 million, were well aware of the disagreements among computational chemists and insisted on a strict reevaluation only a few years later. In 1980, they decided to shut down the NRCC. A report summarizes: "Plug Pulled on Chemistry Computing Center" (Robinson 1980). One particularly significant reason for the shutdown was that new super minicomputers had become available in the 1970s. Their performance was comparable to the performance of older supercomputers, but they could be afforded, housed, maintained, and controlled by a single department. Hence, compared to supercomputing centers, minicomputers offered better possibilities for exploratory modeling. Bolcer and Hermann (1994) aptly summarize that the NRCC lost its significance just at the moment it was implemented.

Another bottleneck for prediction was software. The dynamics of QC depended on software as much as on hardware. Performance of the machine is not the only criterion. The usability of a software, including problems with infrastructure and distribution, are of at least equal importance for modelers.[34] Not all software is open source, meaning that users may not have access to the actual program code. Even if they do, it may remain difficult to actually "understand" how the program works. A second, additional challenge was the compatibility between different platforms or types of machines that were in use (e.g., the IBM 7094, CDC 3300, TR, or IBM 360). In general, the question of portability of code was (and still is) a nontrivial problem that had to be addressed when one intended to work on different machines or to distribute the code to other researchers. A third obstacle was how to organize the distribution of programs—a question that directly affected the identity and progress of QC.

The standardization of software had a significant effect on the evolving identity of QC.[35] A prime example is Gaussian, a software package first released in 1970 that was a comprehensive suite of quantum chemical models developed under the leadership of John Pople (1925–2004) at Carnegie Mellon. He assembled a "club" of experts who contributed various modules to the package with the clear goal of addressing the broader audience of chemists who did not want to develop computational models but rather wanted to use these models. The package is named after the computationally convenient and efficient Gaussian orbitals and Gaussian basis functions mentioned earlier, whose employment served as a kind of computational rationale for the package.

Pople advocated standardization—an arguably inevitable process to foster a growing field. He promoted the use of standardized test beds to be included in the programs' extensive database about well-researched substances that he called "model chemistries." The performance of any model could be tested—computationally—by comparing its predictions with existing model cases. "Configuration interaction" was one example of the way standardization directed research. "Configuration interaction" was a method proposed in the mid-1950s, but it was too computationally demanding until standardized by Gaussian-N model chemistries, when it became a ready-to-use tool of known computational order $O(N^5)$.[36] In addition, the coupled cluster method of the late 1960s was infamous for using formal mathematical manipulations but became popular once these manipulations were implemented—and thereby removed from the user—into the Gaussian suite.

Computational infrastructure changed again in the early 1980s when even smaller, affordable computers came onto the market. One of the reasons for the success of MOPAC, the general molecular orbital package, was that it capitalized on these new computers—namely, the VAX 11/780 minisupercomputers from Digital Equipment Corporation. These machines could be set up in a lab and had output screens so that researchers could use them interactively—an opportunity that greatly increased the popularity of software such as MOPAC. The distribution of software was connected increasingly to the distribution of new hardware. In their brief memoir, Lipkowitz and Boyd (2000) saw the VAX 11/780 as a forerunner of the move to desktop computers operating on a departmental or even personal scale. They wrote, "These machines changed significantly the way computational chemistry was being done at that time and expanded the horizon of computing for many chemistry departments" (viii).

By then, the field of computational chemistry had grown so much that very different levels of expertise coexisted. The popularity of MOPAC, for instance, was based on the fact that one did not need to be an expert on computation and programming to use it. In general, this marks an important development: groups of users outstripped developers and increasingly constituted nonoverlapping communities. Researchers who used computational models to derive predictions about molecules or substances of interest for them were not usually specialists in computational strategies. This presents us with a new social organization of QC—arguably a transformation typical for many "computational" fields. This new organizational structure is not of an exclusively social nature. Rather, it is based on a combination of developments in hardware technology, in software and computational modeling, and in social organization. The question then is: Does this transformation lead to a new culture of prediction? And if so, how does it differ from the iterative–numerical culture linked to the mainframe computer? Before we turn to this question, we shall discuss how the iterative–numerical culture reshaped a central concept—namely, ab initio prediction.

4.7 Interlude: Ab Initio, Prediction, and Complete Prediction

In the context of the iterative–numerical culture of prediction, the term ab initio received a new meaning, and ab initio methods became popular or even hegemonial in QC. This conceptual shift tells something seminal about the iterative–numerical culture of prediction. Though the term ab initio was originally used in computational chemistry in the 1950s, the notion became an important marker in QC around 1970. The Latin term *ab initio* can be translated as "from the beginning" and has many usages in scientific as well as extrascientific domains. In science, it is often used synonymously with "from first principles." When the term was coined in QC, computing from the beginning meant computing from the Schrödinger equation. The first appearance of the term ab initio was in Parr et al. (1950), and it was quickly and widely adopted in the QC community to denote an opposition to semiempirical methods. Of course, the latter also relied on computation, but—in contrast to ab initio methods—they not only referred to laws and natural constants but also depended on the insertion of other empirically obtained values of physical significance.

In this section, we want to bring out a certain ambiguity in the conception—and also in the usage of—ab initio.[37] This ambiguity arises from the fact that "opposing semiempirical" does not coincide with "being derived from first principles." Rather, these two properties belong to different dimensions. When Parr et al. brought up the term ab initio modeling, it was the computational aspect that was key. In their paper on configuration interaction in benzene (1950), their point was that the computation could run from start to finish without the interventional insertion of empirically obtained values. Here, ab initio is used in the sense of automation: the goal was to specify a model, implement it as a program, and then assign particular initial conditions to this program as input to the computation. From this point onward, the result would be produced automatically without further intervention. This kind of procedure was also called "complete prediction" to emphasize the difference from semiempirical methods in which computation requires empirical values to be input at various stages.[38]

Parr and his colleagues' goal was to compute chemical properties rather than measure them. The electronic digital computer was exactly the instrument that let the automation strategies get off the ground. We already discussed Ransil's work on nitrogen and Mulliken and Roothaan's claims about breaking the computational bottleneck. With the first computational models running without any intermediate insertion of empirically determined values, some quantum chemists such as Mulliken (Parr was in his group, too) became strong advocates of computer use in QC. Quickly, the term ab initio was used to discern different camps in QC.

Already in 1960, Coulson had delineated the split of the field of quantum chemists into "ab-initio-ists"—whom he described as "electronic computers" (i.e., as people counting on automation by computers) and "a-posteriori-ists," which, for Coulson, meant "nonelectronic computers" (i.e., advocates of mixed, semiempirical strategies; Coulson 1960, 170). Apparently, the appeal to ab initio as a category was more important than finding elegant labels for the two groups. His statement documents that the term ab initio was already so common that it could indicate demarcation and also that the central issue consisted of electronic computation. The attempt to automate is aptly described as aiming for "complete prediction." By using the term ab initio, however, one suggests that these methods would derive their predictions from first principles. However, this is not the case because of the nature of computational modeling.

Parr's configuration interaction, for instance, can be seen as an analog of Hartree's self-consistent field (SCF) method, discussed earlier in this chapter, that relies on an iterative computational strategy. The outcome, or termination, of SCF depends on a criterion defined by the results of the computation itself: becoming stationary or "self-consistent." Hence it is an instance of "full prediction" or automation but also relies on ad hoc assumptions about its initial values as well as computational strategies that are justified by performance rather than by being derived from first principles.

Thus, we can discern two different senses of ab initio: one is the *principled* sense in which ab initio computation means compute (solely) from the Schrödinger equation. The other is the *automation* (or computational) sense—that is, achieving complete predictions—that tolerates computational models that do not rest on quantum theoretical considerations. Although some parts of the reputation of ab initio methods may well rest on the principled understanding of this term, our argumentation suggests that computational modeling normally relies on the second, computational sense. For instance, consider a set of basis functions. The set determines the class of functions out of which an element with optimal fit can be calculated. This choice will normally be shaped by reasons of computational tractability—theoretical reasons, of course, but not quantum theoretical ones.[39]

An oversimplified view of this issue can be misleading. In their "broken bottleneck" paper, Mulliken and Roothaan (1959) compared their recent development of QC—that is, the first successes in automation and complete prediction—with the application of Newtonian mechanics to engineering. In both cases, it had taken many years to find efficient mathematical formulations to treat problems quantitatively. Now, the authors claimed, this time has come in QC. Hence, they suggested that QC was now able to sail in the same waters as rational mechanics. However, there are reasons to disagree with their standpoint and analogy. Historical and philosophical literature has shown convincingly that engineering knowledge is quite different from an application of theoretical scientific knowledge. In particular, it may be the pretension of rational mechanics that a quantitative treatment of practical problems can be derived from it. But this pretension is at odds with the historical record, as many of our earlier cases in chapters 2 and 3 have shown. Successful predictions are regularly based on sophisticated modeling. When these models refer to principled theory, this does not mean that these theories would in any sense imply the predictions. When Mulliken and Roothaan took the success of computational-ab-initio approaches (complete prediction) as an indicator for the power of principled-ab-initio approaches, they fell into a similar trap.

Ignoring the difference between computational ab initio and first-principle ab initio suggests that the success of the computer has made true the old rational–mechanical dream of first-principle-based derivation. Such a viewpoint neglects the significance of computational modeling that is not an issue of theoretical derivation but rather an issue of mediation between theory, experiment, phenomena, and computational technology.[40] It may be especially important to emphasize that technology also influenced QC on the experimental side. Whereas new computational instrumentation did lead to an upswing of ab initio approaches, new instrumental, experimental, and laboratory technologies (e.g., spectroscopy) also made available new and much more refined data.

The story of QC began with two different possible approaches: the principled one and the semiempirical one. It is true that the semiempirical one was the leading approach, whereas the principled one more or less got stuck in an impasse early on. However, from the time that ab initio was coined in the early 1950s, it marked an opposition to semiempirical strategies. Back then, a majority of chemists saw the ambitions for sophisticated computational modeling as neither necessary nor helpful because they narrowed the prospects for obtaining chemically relevant results. In the 1970s, ab initio methods started to become predictive, and, a decade later, a substantially broadened field of applications was much more easily accessible for computational chemists. By that time, ab initio (in the *computational* sense) had become the leading approach in the community of computational quantum chemists.

John Pople's career illustrates the point. He was always deeply interested in mathematical modeling, but it was clear to him that mathematical and computational approaches should have an auxiliary role in chemistry. He thought that QC should focus on cases of chemical interest. Because this was hardly possible with ab initio methods, Pople preferred a semiempirical approach.[41] In the 1970s, however, increasing computational power and the development of standard computational methods made it possible to tackle interesting cases on an ab initio basis. Pople became an influential proponent of ab initio methods, not least by codeveloping and promoting the Gaussian software package.[42]

However, it is necessary to distinguish the fact that a mathematical *prediction* was possible from the existence of a mathematico-theoretical *foundation* of chemistry. The increasing range of predictive computational modeling in chemically relevant cases does not per se justify a strong foundational claim.[43] Rather, it proves the growing efficiency of ab initio *in the computational sense*. This marks a central point: computational modeling requires measures that have neither a principled nor a semiempirical character. A principled strategy would have to specify a model completely from quantum theoretical considerations. A semiempirical strategy—in the sense established in QC—would plug in empirically obtained values of physical significance. Separate from both, computational models follow their own dynamics, sensitive to formal and instrumental conditions alike. Thus, the term ab initio exemplifies the autonomous dynamics of mathematization insofar as it adopted a new meaning: in the rational culture of prediction, "from the beginning" meant to predict by derivation from fundamental laws, whereas in the iterative–numerical culture, the term adopted a computational meaning.

4.8 Two Gambits and a New Era: Density Functional Theory

The iterative–numerical culture of prediction was fully established by the 1970s. But the story does not end here. In the 1990s, a new culture developed that (again) shaped QC. We shall observe and analyze this culture by tracking the history of density functional theory (DFT). DFT actually has a longer history in solid state physics in which it has played an important role since the mid-1960s, but it gained outstanding relevance in chemistry with the turn to "small"—that is, highly available and networked—computers. This turn indicates a decisive inflection point in the epistemology of iteration, leading to a new culture of prediction that we call exploratory–iterative.

One would expect an upswing of computational QC on the grounds of the general availability of computational power. DFT, however, stands out: "The truly spectacular development in this new quantum chemical era is density functional theory (DFT)" (Barden and Schaefer 2000, 1415). One can readily read off the "spectacular development" from bibliometric data. A simple survey shows that the number of articles with DFT in the title or abstract started off like a rocket in the decade of the 1990s, quickly rising to a staggering fifteen thousand papers per year in 2015 (see figure 4.1).

Thus, the obvious question is: What happened with DFT around 1990? DFT was not a new theory; it had originated in solid state physics thirty years earlier. How and why did it become relevant in chemistry more than a generation later? One answer to this question focuses on the development and availability of new computational instruments. Another highlights a new conception of computational modeling. Still a third hinges on a reorganization of the field, even the discipline, of chemistry. Together, these factors make up a new configuration.[44] More specifically, we claim that DFT became a shining example of a new culture of prediction.

To explain and justify this claim, we first have to introduce DFT. We shall keep the necessary technical arguments to a minimum. QC generally deals with the electronic structure of atoms and molecules. The Schrödinger equation expresses the energy of such structures as a wave equation. Attempts to solve this equation had been at the root of QC since 1927,

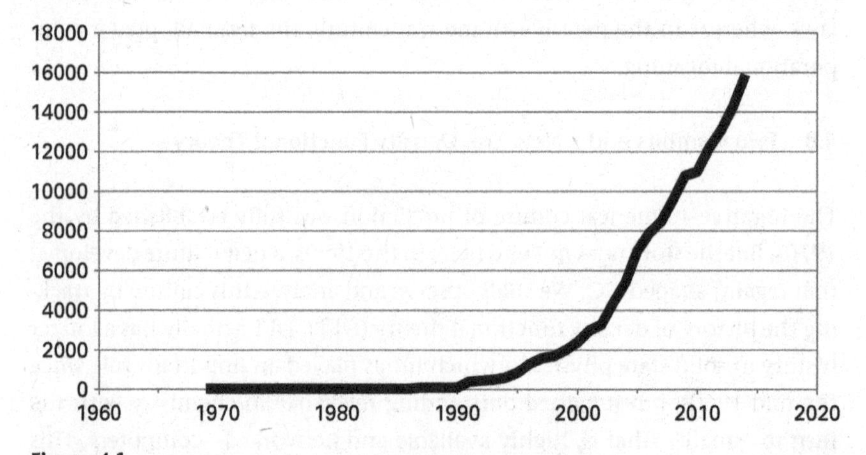

Figure 4.1
The number of papers with "density functional theory" in the title or abstract per year in ISI Web of Science database. Courtesy: authors.

and dealing with the extraordinary complexities of the equation shaped QC into an iterative–numerical culture. DFT addresses the same energy as the Schrödinger equation does, but in a mathematically alternative way that promises to remove much of the complexity. Due to electron interactions, each Schrödinger wave function has $3N$ degrees of freedom (N being the number of electrons)—a large number. DFT, however, expresses the same energy in terms of the joint electron density—that is, an "object" in space with only three degrees of freedom (see figure 4.2 for an illustration).

The computational advantages of reducing complexity have led to the practical use of this approach in engineering, but the legitimacy or theory behind this approach was not clear; it held a more heuristic status. The condensed matter physicist Walter Kohn (1923–2016) played a major part in advancing this heuristic approach to the level of theory. He and his colleague Pierre Hohenberg (1934–2017) produced two theorems (Hohenberg and Kohn 1964). Their first theorem states that the ground state energy is determined uniquely by the corresponding electron density $\rho(r)$, that is, $E = E\,(\rho(r))$. This equation has to be read as saying that the energy E can be expressed as a function only of the electron density ρ. The second theorem is a mathematical variational principle: $E\,[\rho_{trial}] \geq E\,[\rho]$—that is, any hypothetical density will give a larger energy than the correct one, and hence: "the exact ground state energy and density can be calculated without recourse to the Schrödinger equation, at least in principle" (Bickelhaupt and Baerends 2000, 3). Therefore, DFT determines the energy without recourse to the Schrödinger equation and its devastating complexity. However, as Bickelhaupt and Baerends aptly write, the promise holds "in principle,"

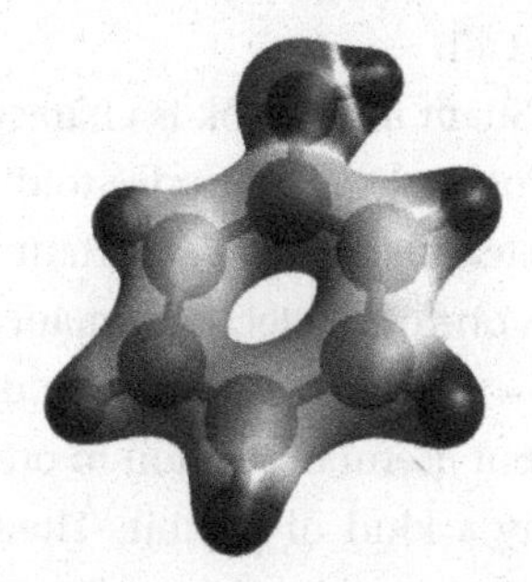

Figure 4.2
The electron density of phenol. The darker regions refer to higher density (i.e., the probability of electrons visiting this region). Courtesy: authors.

whereas *in practice*, there are still serious problems of computational complexity. Hohenberg and Kohn proved the existence of a function that gives the energy and depends only on the electron density. But this existence is meant in a mathematical sense—that is, the sheer existence of such a functional relationship between E and ρ is proved, but the theorem does not give a clue as to what that function looks like or how it can be determined.[45] The interaction of electrons that we already identified as a main source of computational complexity—electron exchange and correlation effects—is covered by DFT in an implicit way by the (unknown) functional relationship itself. The space of mathematical functions is extremely large. Hence, actually determining one particular function might be very difficult. Up to this point, the elegant and mathematically proven theorem did not show a way of predicting. It was like an elegant suitcase without a handle.

Kohn was aware of this shortcoming and, together with his coworker Lu Jeu Sham, he introduced a practical computational scheme (Kohn and Sham 1965). This scheme postulated a reference system of N noninteracting electrons, so that interaction effects (exchange and correlation of electrons) can be captured by a local potential $v_{xc}(r)$—a deliberately counterfactual assumption. The modelers assumed—in favor of the model—that the exchange and correlation effects can be expressed via the local potential to a sufficiently appropriate degree. At the same time, this assumption is a crucial idealization because it opens up paths for computational treatment. The (hypothetical) Kohn–Sham potential was an attempt to deal with the unknown functional relationship by (counterfactually) assuming an idealized situation. It places a numerical handle on the problem of how to approximate the unknown functional and has been the main basis for further developments in DFT.

The move Kohn and Sham undertook is characteristic of mathematical modeling; variants of it occur in all episodes told in this book. It aims to reconcile prediction and feasibility. It is important to realize that the question of appropriateness is one of model performance, not one of theoretical justification. The model—the local potential—is deliberately false in relation to quantum theory but useful in relation to prediction. In other words, resourceful modelers play a kind of gambit. They reduce the bond with theory and create a model that can make predictions efficiently. What efficiently means depends on what instruments are available, and using these instruments requires adequate mathematization. Therefore, the gambit has to be reinvented by each culture of prediction. Of course, like any gambit, the loss is certain and the gain not guaranteed.

DFT was immediately successful in solid physics, particularly in crystallography in which molecular structures are rather regular. Kohn's 1964 and 1965 publications (with Hohenberg and with Sham) were immensely influential papers. Indeed, they are the most highly cited papers ever in the *American Physical Society*'s flagship journal *Physical Review* (Redner 2004).[46] However, DFT had almost no uptake in chemistry. The number of papers on DFT in chemistry journals languished at around thirty per year throughout the 1970s and 1980s. The Kohn–Sham approximation scheme did not provide predictions accurate enough for the less regularly structured cases that were the main interest for chemists.

All this changed fundamentally in the 1990s. DFT shot up to become the arguably most highly used theory in all of science. A large share of the 15,000 articles (in 2015 alone) comes from chemistry, which documents that the "spectacle" Bickelhaupt and Baerends diagnosed in 2000 is still ongoing. Another indicator of DFT's prominence is the Nobel prize that Kohn won in 1998 "for his development of density functional theory." To his own surprise, the theoretical physicist Kohn received it in chemistry. He shared the prize with John Pople, the mathematically minded chemist and organizer of Gaussian, who earned it "for his development of computational methods in quantum chemistry." By the way, this was the first time in its history that the Nobel Prize was awarded for computational modeling. Somewhat ironically, before the turn, the Gaussian software package had not included DFT approaches because they were not seen as competitive in relevant cases.

The 1990s turn rests on a new conception of computational modeling that synthesized two elements: first, a strategy of mathematization that assigns exploration a pivotal role, and second, the technology of easily accessible and networked computers. In a sense, this new conception further expands the autonomy of modeling.

We have already seen that models comprise artificial and deliberately false components, as illustrated by the Kohn–Sham scheme. The local density approximation is a theoretically informed parameterization scheme that allows for different particular specifications, such as the exact form of the local potential, that are not derived from quantum theory. However, such parameterization influences the performance so that the model can be adapted to known experimental data or model chemistries. This is done through iterative procedures that check performance, then modify the model or the parameter values, then check again, and so on.

This type of approach is typical for computational modeling (see Lenhard 2019, chap. 1), and it provoked an equally typical criticism because artificial components seem to lack justification. In our current example, the assumption of the Kohn–Sham approach is that for every system of interacting electrons moving in an external potential, there exists a local potential such that a system of noninteracting electrons will obtain the same density. The status of these assumptions was debated. "For some time a physical meaning of these KS orbitals has been denied" (Bickelhaupt and Baerends 2000, 5). The crucial question then is: Can predictive performance compensate for a deficit in physical meaning? A culture of *prediction* allows an affirmative answer: the gambit of mathematical modeling.

The 1990s turn comes with a further change in perspective. Waiving physical meaning (partly) ceases to be perceived as a move that needs substantial justification. Instead, constructing models with many adjustable—that is, not yet specified—parameters becomes the standard. Although such models are still theoretically informed, the increasingly important role of parameterizations makes the model's performance in prediction all the more important. Mathematization fulfills a new and distinct function here. *It provides the stage for active exploration.* The perceived drawback—this is what the talk of gambit concedes—became a favored move. In this way, the 1990s turn adds a second gambit to the first one.

This change in the conception of modeling rests on a technological component as well. Exploration must be practical—that is, using the computer to explore the behavior of a seriously underspecified model must be a feasible option for researchers. There is no other way to handle "artificial" parameters. Because they often lack physical meaning, they have to be justified by performance; and, moreover, their values have to be assigned via an *exploratory* procedure. Such procedures, in turn, require constant access to and direct feedback from computers. This was provided by the powerful but affordable, easily accessed, typically networked computers that became widely available during the 1990s. Only then did a new combination of in-principle theory complemented by a layer of iterative, semiempirical adaptation become attractive and the basis for an exploratory mode of research.[47]

In the late 1980s and early 1990s, Becke, Lee-Yang-Parr, and Perdew were among those who introduced a new generation of density functionals. These functionals contain a relatively large number of adjustable parameters. Moreover, most of these parameters lack physical meaning. At the same time, these functionals work for model chemistries—that is, they perform well in the standard cases of QC. This development exemplifies the new type of mathematization. In a sense, modeling becomes more formal insofar as it is guided by mathematical properties (of the parameterized functional) rather than physical meaning. These mathematical properties, however, cannot be examined analytically but only by exploration. The synthesis between modeling and technology became a cornerstone of a new exploratory–iterative culture of prediction that adds an exploratory element to the older iterative–numerical culture.

4.9 Coming to Grips with the New Culture

The epistemology of iteration and the social organization of the new culture coevolved. Both users and functionals multiplied. DFT quickly diversified into a host of various functionals, and they did so in two dimensions: first, via parameter assignment, one functional can be adapted to specific conditions, substances, or mixtures of the two. Second, new functionals can be spawned with relative ease. So-called hybrid functionals illustrate the point. Modelers combine existing functionals—for instance, by using a weighted average of them. Which weighing factors work best depends on the case under consideration and can be determined by iterative–exploratory testing of a model against available data. The result is a large number of functionals, most of them streamlined to predict in a specific and probably narrow range of cases.

How a more general-purpose software should handle this situation is not straightforward. Gaussian, to stick with our example, began to incorporate DFT modules in the early 1990s due to the pressure of DFT's predictive success and demand in the community. At this point, the ab-initio-oriented community that developed Gaussian signaled that DFT had been accepted into the acknowledged set of methods. The main argument in favor of DFT was the ratio of speed—that is, relatively low order of computational cost—to predictive accuracy. To the extent that users could access computational resources easily and cheaply, the need to adapt functionals to specific cases of interest did not constitute a drawback.

But if the exploratory part is particularly important, does that not contradict the somewhat unifying and standardizing outlook of the Gaussian software package? In fact, over the past generation, Gaussian has lost much of its standardizing rationale. On the one hand, it has many competitors that offer a diverse range of functionals. A preliminary search on the internet indicates that there are more than one hundred DFT-related packages on the market, of which roughly 25 percent are open, 35 percent academic, and 40 percent commercial. Such packages proliferate because they often specialize in niches defined by certain materials or contexts of application.

On the other hand, Gaussian itself has reacted to the exploratory nature of DFT. It offers an increasing diversity of functionals to users. If one functional is known to work well for one class of substance and another functional for another substance, it may well be that the weighted average (so-called hybrid) is a good compromise when substances are mixed. Accordingly, recent releases of Gaussian provide more than a dozen density functionals, and the user's reference guide recommends trying several of them to cross-check results (Foresman and Frisch 1993).

That even developers put only limited faith in the validity of functionals did not go unnoticed. A main argument against DFT is that, from a theoretical point of view, functionals remained ad hoc insofar as they were in an unclear relation to the correct functional (cf. Barden and Schaefer 2000). Exploration does obscure this relationship, and thus runs counter to a well-established philosophical account of modeling that describes mathematical modeling as building a hierarchy in which one idealization approximates a less idealized level with reality at one end. According to this account, one can move up and down (de-)idealization like a ladder.[48] But such a ladder presupposes a clear approximation relationship. Alas, despite continuing efforts toward clarification, results are meager. When Perdew et al. (2005) reviewed this work—and contributed to it—they bemoaned that prediction and exploration drive the dynamics of DFT and move at a higher speed than attempts at clarification and validation. Recently, the validation problem has been tackled in an attempt that is a true offspring of the exploratory–iterative culture. Lejaeghere et al. (2016) tested the consistency of various DFT softwares not by analyzing their functional form but by a massive parallel test on identical cases (with sixty-eight coauthors)—a computational way of living with the unclear relationship.

The social organization in the new exploratory–iterative culture of prediction differs remarkably from the older mainframe culture. The software suite Gaussian serves as an illustration. It was first released in 1970 as an academic endeavor by Pople and others. Its appearance indicated that QC had settled on a framework and methods. The program itself fostered standardization via the models, algorithms, and model chemistries it implemented. Gaussian remains the market leader in QC software, but around 1990, it underwent significant changes to maintain its position. First, Gaussian started to also provide versions that run on the Microsoft-DOS platform, reflecting that the new field was ready to address users of small, desktop computer systems. Second, in 1987, Gaussian went commercial—that is, it changed from the academic setting at Carnegie Mellon to become *Gaussian, Inc.* From then on, the software addressed a broadened audience of theoreticians, experimentalists, and engineers. Typically, the users of such instruments would not be part of the developer community (e.g., not mathematically well-versed quantum chemists) and would have very different knowledge, skills, scientific goals, and types of projects.[49]

The extraordinarily high number of publications on DFT methods goes far beyond the size of the community of quantum theory experts. Rather, it draws from the much larger reservoir of authors including engineers who are interested in predicting the behavior of molecules. Scientists who work with DFT usually have considerable expertise in finding the right way to work with and to modify functionals. It is exactly the goal of the software to make it easy for the user to work with functionals without being an expert on their derivation or construction. Typically, laboratories and working groups maintain their own inventory of functionals and adaptation techniques plus internal preferences and practices regarding how and when to use them—something new members become encultured to.

The community of DFT users is based on a multitude of distributed local adaptations, all connected through a networked infrastructure. It is the dynamic combination and adaptation of building blocks, facilitated by networked infrastructure, that instantiates the recent success of DFT, rather than the unifying theoretical framework and the quest for something like "the" correct functional. Users do not need great programming skills either. Often, they may not even be able to look at the code because the software code is closed and proprietary (as is the case with Gaussian).[50] Consequently, the emergence of a commercial market and wide accessibility have been accompanied by a loss of expertise on the side of many users. Critical reviews of this phenomenon are legion and by no means restricted to QC. To give just one example, the "tutorial on post-Hartree–Fock methods" warns: "These days, software vendors make it easy to run a calculation at the 'touch of a button.' Is the button you are pushing the right one?" (Lipkowitz and Boyd 1994, vii). In other words, software is part if the social, philosophical, and technological fabric of the new culture of prediction.[51]

Thus, the case of QC since the 1920s spans different cultures of prediction. In the pioneering phase of QC, the rational approach competed with the (semi) empirical one that we also examined in earlier chapters. Whereas the first saw predictions as the natural outcome of theory, the latter picked up the experimental tradition of chemistry. The digital computer brought a turn to an iterative–numerical culture of prediction. This culture included the use of large and expensive (mainframe) computing machines and their centralized organization. A further turn occurred around 1990 when easily and cheaply accessible networked computers became available, and when mathematization moved to a genuinely exploratory mode. An exploratory–iterative mode of computational modeling, we have argued, characterizes this culture.[52]

We conclude our investigation of the exploratory–iterative culture with a brief look at what happened to the concept of the ab initio method. DFT undermines this concept; or, to put it better, the exploratory mode of mathematization undermines this concept. On first view, DFT is an ab initio method in our terminology introduced previously—ab initio in the principled sense. Kohn–Sham's claims about an alternative to the Schrödinger equation are mathematically proven. Being more ab initio is impossible. However, an exploratory mode of modeling is orthogonal to a theoretical foundation, or at least to the idea that the models should be derived from these foundations. We have already discussed the ambiguity of ab initio methods between a first-principle sense and a complete-prediction sense that emerged with computational modeling in the mainframe era. At that point, iteration became a crucial tactic within computational strategies, and ab initio was understood as complete prediction. In the 1990s, an exploratory mindset developed, again giving cause for concern on the side of researchers understanding themselves as ab-initio-ists.

> Historically, a chief tenet of quantum chemistry has been that predictions about the structure and properties of molecules ought to be based on the quantum theory and not on parameterizations, heuristics, or empirical correlations, however accurate they may appear to be. That sentiment has changed quite a bit as the problems have become more complicated. (Barden and Schaefer 2000, 1407)

In other words, the predictions by DFT rely on exactly that kind of measures that ab-initio-ists tried to ban. Complete prediction does not just dilute or weaken the original, principled sense. Even worse, in the exploratory mode of modeling, prediction thrives by violating the principled sense of ab initio. Mathematization is not assuring coherence—as the host of density functionals shows so vividly. Rather, it is managing plurality and opening up ways to diversify functionals and models in the hunt for accurate predictions.

III The Iterative–Numerical Mode

5 Systems Thinking and the Limits to Growth

There has hardly ever been a more dramatic staging for the publication of a scientific study than that of *The Limits to Growth* (Meadows et al. 1972). Commissioned by the Club of Rome, it was presented to a hand-picked audience of scientists and politicians assembled in the Smithsonian Institution. The message was alarming: if the growth of the economy, pollution, and population continues, the world system will collapse in less than a century. News spread in almost no time and informed a broad public audience about the predicament of humankind.

In a sense, this kind of message was nothing new. Warnings about doomsday testify how predictions about the world's fate are almost as old as history itself. Nonetheless, *The Limits to Growth* was different because it was a *scientific* prediction, and one that used a *computer* model. On the political side, it is often identified as the starting signal for an environmental movement that synthesizes political action and scientific prediction.[1] On the scientific side, the study marks the culmination of *system dynamics*, pioneered by engineer Jay Forrester (1918–2016), inspired by modern computers, and claiming to be in possession of the right method for dealing with complexity.

The study was a product of a new type of mathematical modeling. While maintaining the goal of prediction, it geared mathematics toward iteration and complexity. At the same time, the study shows how thinking about the future was being channeled toward computer modeling and advancing under the umbrella of a mainframe culture of prediction. In fact, we claim that *The Limits to Growth* is an exemplar of a newly emerging culture of prediction that has both an iterative and a numerical character.[2]

Our work profits much from the history of computing that revolves around the digital mainframe computer and its rich technological, social, and institutional context. Machines such as the ENIAC and conceptual documents such as the "First Draft of the EDVAC" were groundbreaking achievements that are widely acknowledged by historical scholarship. This scholarship has made it clear that computers did not originate from there simply in a straight line but rather have prehistories showing how they developed along a branching network.[3] In this chapter, we concentrate on Forrester's work because it is a critical piece of history that shows, as if under a magnifying glass, how technology, mathematical tools for making predictions, institutional contexts, and conceptions of the future have *coevolved*.[4]

The following section 5.1 starts with the prehistory of *The Limits to Growth*. It examines Forrester's work at the MIT as a leader of Whirlwind, a project that started out with the mission to manufacture a pilot training device and ended up building a real-time feedback computer. It was the fastest and most expensive early generation digital computer that went operational in 1952 and became part of the air defense system SAGE. Typical for mainframe machines, Whirlwind was an expensive resource embedded in an institution that organized its use and controlled access.[5] At the same time, Whirlwind is a special case because Forrester and his coworkers arrived at the iterative–numerical mode of prediction via an engineering approach that was in competition with the mathematical one. Furthermore, we argue, Whirlwind's design informed Forrester's conception of system dynamics. He was hunting for a sort of holy grail: a techno-scientific system that could inform decisions via computer-based predictions in a very general—and very technocratic—way.

Section 5.2 examines Forrester's system dynamics, an instance in a broader movement of systems thinking that became popular during and after World War II when it "effloresced into a number of forms, including operations research, systems engineering, systems analysis, and system dynamics" (Hughes and Hughes 2000, 1). Each of these variants paved a new path to predictions, and each attributed an important role to the computer.[6] Forrester ignored existing mathematical techniques almost completely and geared system dynamics toward the computer in a way that made modeling look almost effortless. Indeed, while the Club of Rome was pondering whether modeling the entire world system would be feasible, Forrester created his model in just a few hours. For him, the step from a company to a city or even to the world required no more than some minor adaptations in the model.

The Limits to Growth did not just attract wide attention; it also drew criticism. Section 5.3 discusses various lines of critique that demonstrate the problems with which the new culture had to contend. In particular, the study's validity was questioned from a mathematical perspective, from the (quickly evolving) computer modeling community, and from the established culture of expertise. At issue was to what extent the new culture of prediction could establish its own standards for validity.

The Limits to Growth marks a turn after which thinking about the future was perceived increasingly as an activity inside the mainframe culture of prediction. Section 5.4 explores a wider perspective and locates the discussion in the futurism of the 1960s and early 1970s. Does this establish the future as a legitimate scientific subject—or rather eliminate it? We invoke the philosopher Hannah Arendt and the historian Reinhart Koselleck to suggest that conflicting answers are possible.

5.1 Whirlwind: Hunting the Ultimate

Jay Forrester, a young engineer with experience in radar systems, earned his spurs as the manager of the Whirlwind project at MIT.[7] This project started as a mission impossible in 1943–1944 when the navy's Bureau of Aeronautics initiated a plan to develop an *Airplane Stability and Control Analyzer* (ASCA). Unlike existing flight trainers, the ASCA should be a universal device able to simulate any airplane by *computing* the feedback response to the pilot's actions based on a physics model of the aircraft's dynamics. The ASCA was never built. The project followed a winding path, including a change in goals from the ASCA to a general-purpose computer and in funding institution from the navy to the air force in 1950. When Whirlwind went into service in 1952, it was "the first high-speed electronic digital computer able to operate in 'real time'" (Redmond and Smith 1980, vii),[8] and it built the computation backbone of the behemoth SAGE air defense system.[9] This path shaped Forrester's concept of system dynamics.

When Forrester examined existing flight trainers as well as the servomechanical[10] equipment being developed at the time at MIT, he realized that crucial components were not available—the ASCA simulator required a versatile, fast, and reliable computer; a formal mathematical model of the aircraft; and some version of computational fluid dynamics that would make the former two elements work together. In short, the ASCA was "essentially a physicist's dream and an engineer's nightmare."[11]

Forrester had the stamina to not be paralyzed by this nightmare. He first geared the project toward creating a computer, but achieving the desired speed and reliability still looked to be an unsurmountable task. A lucky coincidence helped him out. His colleague Perry O. Crawford, who was well networked in the computing community, suggested to Forrester that some *digital* machine might look promising (see Mindell 2002, 291). In fact, the very first conference on digital computing was about to take place at MIT, the "conference on advanced computing techniques" (October 30–31, 1945). It was a revelation for Forrester. Presper Eckert, John Mauchly, and their team at the Moore School of Electrical Engineering reported on the ENIAC, the digital computer that was about to be presented to the public and was much admired at the conference. Another highlight was John von Neumann's presentation of the EDVAC design that promised a far more versatile machine.[12] Forrester made up his mind immediately: his MIT group needed a digital computer.

However, the principal advantages of such a machine by no means implied that the design was settled. Even if, in theory, a computer had the potential to fulfill the reliability, speed, and versatility demands of the project, in practice, no such technology existed. Even worse, the decision to concentrate on the computer part alienated the navy that had commissioned a training device for pilots. In this situation, Forrester's rhetorical skills proved to be at least as important as the technological and scientific expertise of his group.[13] More than once, he was able to secure more funding (costs were excessive) and avert the impending project termination. The story of the ASCA/Whirlwind is as much a story about institutions and social organization as it is one about scientific and engineering challenges.[14] However, the ASCA prehistory was inscribed into the Whirlwind computer. The machine was designed for high-speed feedback (real-time computation) so that it could act as the central control instance in a larger *system of prediction* in which decisions could be urgent and predictions would have to be delivered immediately. Ideally, so Forrester envisioned, such a system would be flexible and universal enough to deliver predictions in almost any context.

> Were a stranger to ask, "Can you do my computing job?" wrote Forrester, the answer appropriate to a minimum computer's capacities must be, "Probably, but we must analyze it to find out." But if one possessed the ultimate system Forrester had in mind, then the answer to the question could safely be: "Yes, what is it?"

> Forrester was after the ultimate. (Redmond and Smith [1980, 186], citing Forrester to Stratton, March 3, 1950)

The air force took over, and the Whirlwind computer became operational in 1952 (see figure 5.1), then forming the computational core of the SAGE system for air defense. It worked with computer-based real-time analyses of incoming data from radar stations and made very quick decisions about impending threats—supposedly such as Russian aircraft.[15] This system encompassed vast dimensions in terms of its geographical spread, the people involved, and its development costs. A system that marshals amounts of money of the order Whirlwind did will be accessible to only a narrow and highly qualified circle of customers.[16] Even if the computer might be called general-purpose, it was a part of a system targeting very select purposes. In this, Whirlwind resembled other mainframe machines. And this is a typical feature for the mainframe culture of prediction—while computing machines in a somehow ideal sense are called general-purpose, in their factual institutional context, they have extremely restricted purposes. Consequently, predictions, or the attempts to get predictions, are channeled toward such purposes.

Here is an illustration of how precious the Whirlwind computational resource was. Forrester had spotted the expensive and fragile vacuum tubes as the weakest spot in the technology of Whirlwind.[17] His arguably most significant achievement in engineering was the development of magnetic storage. It made computer memory simultaneously faster, much cheaper, and more reliable (see figure 5.2). When he and his group came up with the

Figure 5.1
Whirlwind computer room, 1952. Used and reprinted with permission of The MITRE Corporation © 2023.

Figure 5.2
Tube versus magnetic storage; *top*: the circuitry from Project Whirlwind's core memory unit; *bottom*: the core planes from Project Whirlwind's core memory unit. Displayed at the Charles River Museum of Industry, Waltham, Massachusetts. Attribution: Dpbsmith at English Wikipedia, creative commons.

prototype of the new storage in 1952, they needed to test it on the Whirlwind system to make sure it actually worked as well as the laboratory results promised. However, the machine was already in operation, and it was not possible even for Forrester to get some computing time on Whirlwind. The tasks assigned to it were deemed too urgent to allow any intermission.[18] Forrester's group had to build a separate (smaller) computer to test the storage. The expensive technology of the computer and the urgency of the demands mutually justified each other. Whirlwind had grown into a gigantic high-speed machine. As an example of early mainframe computers, Whirlwind was outstanding in terms of its size, speed, and cost. At the same time, typical for mainframe machines, it was part of an even larger system that regulated access to the instrument.[19]

When developing Whirlwind, Forrester and his group followed a somewhat special concept of mathematization. They started from mechanical

Figure 5.2
(continued)

apparatuses (such as aircraft simulators) and thought of computational means to speed things up. That is quite different from starting with mathematical equations. In other words, they approached high-speed iteration from the engineering side, as mathematization of the servomechanism. That formal mathematics should not have the leading role met with skepticism.[20] In particular, Mina Rees and her colleagues from the mathematics branch of the Office of Naval Research were concerned about what they saw as a lack of mathematical expertise in the Whirlwind group. In her critical reviews, she took as her point of comparison the general-purpose digital computer that von Neumann and Goldstine developed at the IAS, Princeton. Her principal point was that Whirlwind suffered from exploding costs and could achieve little because the key problem was a mathematical-conceptual one on which Forrester's group had insufficient expertise.[21] Forrester needed all his rhetorical skills to save the project. His strategy was to sell Whirlwind not just as a general-purpose computer but as part of something bigger—a predictive *system* supporting real-time *decisions*. In this way, he fashioned prediction as a link between science and policy. And building this link meant designing expensive technology. Forrester and Everett (his coleader) argued that the IAS was building a computer that would "apparently be for their own laboratory studies and mathematical problems" (cited in Redmond and Smith [1980, 124]). The MIT computer, on the other hand, must meet special requirements in terms of speed and reliability that were outside mathematical expertise and made the project so expensive.

Whereas the navy's Office of Naval Research (ONR) remained skeptical, the project was saved financially when the air force took it over in 1950. Nonetheless, the MIT and IAS groups followed different approaches to mathematization. The IAS group targeted established and intractable mathematical problems such as solving systems of partial differential equations, and they viewed the computer as a new instrument to solve these equations with new numerical tools. The Whirlwind group proceeded differently. Indeed, they were not qualified to address the mathematical problems regarding fluid dynamics and the like—Rees was right on this point. Instead, they adopted an engineering perspective and undertook to mathematize the feedback functions of servomechanism by iterative–numerical means.[22] The Whirlwind group more or less assumed that the mathematical concepts were there and that the hardware was the point. Forrester did not feel at all indebted to mathematical problems. For him, the capabilities of the computer came first, and the mathematical part of the predictive system would simply fit in with this. That is, there were different and partly conflicting views regarding what mathematization would look like in the emerging mainframe culture of prediction.

Thinking on prediction coevolved with technology. The mainframe culture of prediction began to flourish in the 1950s, thriving on big science and military funding. Computers were expensive, and access to them was highly regulated. Over the course of the 1960s, computers became more accessible instruments and the mainframe culture broadened.[23] Forrester's work can still serve us as an example. After Whirlwind became operational, he left the project. However, he remained in the orbit of MIT and joined the *Sloan Business School* where he founded the *System Dynamics Group* in 1956. By way of system dynamics, Forrester could pursue his vision of a system of prediction. The following sections examine how *The Limits to Growth* study exemplifies the mainframe culture of prediction.

5.2 Systems Thinking: A Machinery for Prediction

In 1970, the Club of Rome, an informal and exclusive circle that perceived itself as an "invisible college," was preparing to commission a study on the "predicament of mankind." At the time, issues such as the nuclear threat, population growth, and environmental pollution had created a sense of urgency and the belief that society needed to decide about its own future.[24] The Club of Rome thought that computer models were the right means for predicting possible futures and finding out how a transition from growth to a (noncritical) equilibrium state might happen.[25] The Club asked Forrester, the famed pioneer of computer-based management models, to inquire whether a study was doable. Forrester happily accepted. In an unforeseen way, a stranger had asked for a prediction—what could be a more challenging task for Forrester than predicting the future course of the world?

When the Club of Rome asked Forrester, something remarkable happened. It took him only a couple of hours to set up the model (see Forrester 1971). System dynamics apparently had been waiting for the Club to ask. This observation, we argue, tells how much systems thinking[26] is a machinery for prediction. Of course, the almost immediate answer the Club of Rome received from Forrester was conditioned on buying into the model architecture of system dynamics.

The full study was conducted by MIT's System Dynamics Group led by Forrester's former student Dennis Meadows with Forrester acting as consultant (Edwards 2000, 243). Forrester was not shy about showcasing the role of his approach. In 1971, before *The Limits to Growth* (*LtG*) came out in 1972, he managed to publish his own study *World Dynamics*. In the introduction, he claimed ownership of world modeling.[27] Both Forrester's study and *LtG* promoted the systems framework as a new way of perceiving the world.[28] In fact, there was allegedly no alternative:

> All [members] are united . . . by their overriding conviction that the major problems facing mankind are of such complexity and are so interrelated that traditional institutions and policies are no longer able to cope with them, nor even to come to grips with their full content. (Watts in Foreword to Meadows et al. [1972, 9/10])

This reads as if the insights into complexity and political urgency force the move to system dynamics and computer technology—a message conveyed consistently in both Forrester's and Meadows et al.'s studies. However, there is a rich texture of values and preferences that are part and parcel of system dynamics.

For Forrester, the world constituted but one instance for the system dynamics approach. According to this, the structure of any system is defined by entities and flows between them that are controlled by valves. The flows depend on the state of the entity, thus creating a system of feedback interactions much like in a servomechanism. The overall dynamics result from the mutual influence between entities that regularly takes the form of nonlinear relationships (because feedback effects accumulate). This property is infamous in mathematics because a solution—that is, determining the system behavior by integrating basic equations—becomes intractable.[29] However, this is of little concern in system dynamics because the system behavior is computed forward step by step, thriving on iterative capacity while avoiding intricacies of mathematical analysis. This kind of algorithmic strategy needs a computer to iterate the feedback over and over until some behavior becomes manifest.

Forrester himself identifies four pillars on which system dynamics rests: (1) information-feedback control theory inspired by servomechanisms, (2) decision-making processes as a task for prediction adapted from military tactics, (3) replacing mathematical analysis by computer experimentation, and (4) digital computers whose cost had fallen so much that they became economically feasible for users beyond the military.[30]

Forrester's first book was on *Industrial Dynamics* (1961). It sets out to reduce all functional areas of managing a company to a common basis

> by recognizing that any economic or corporate activity consists of flows of money, orders, materials, personnel, and capital equipment. These five flows are integrated by an information network. . . . Industrial dynamics is a way of studying the behavior of industrial systems to show how policies, decisions, structure, and delays are interrelated to influence growth and stability. (Forrester 1961, vii)

One pertinent example from this book is the supply chain of the beverage industry and dealers. The stock in soda or beer depends on feedback loops between demand and planning. These can easily lead to an unstable situation in which the stock shows overshoot and collapse—the signature phenomenon of system dynamics (see figure 5.3).

The model architecture is quite generic. Renaming the entities and adjusting their interaction handily results in another subject such as urban dynamics.[31] Beer shortage in a retailer's stock is not so different from food

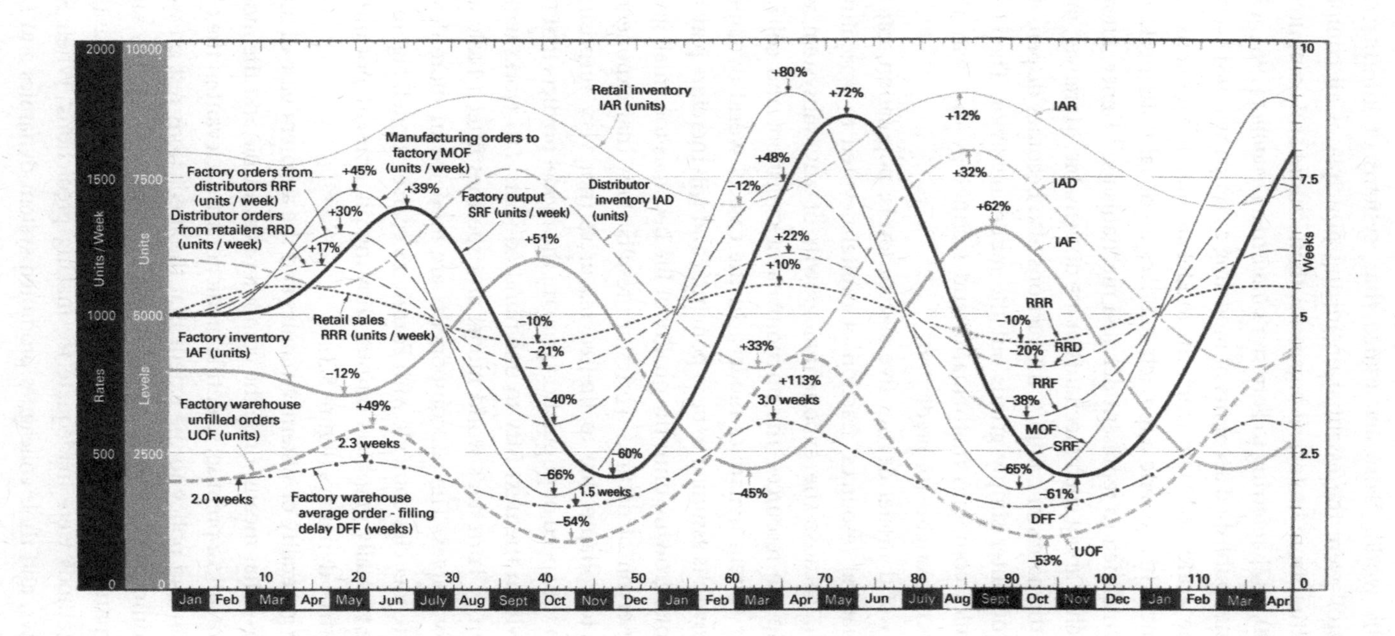

Figure 5.3
The response of a production-distribution system to a 10 percent unexpected rise and fall in retail sales over a one-year period (Forrester 1961, 26–27).

shortage in the world. "Forrester argued that 'systems of information-feedback control' were the essential organizing principle of all complex organized entities, from biological organisms to machines and computers" (Edwards [2000, 237], referring to Forrester's *Industrial Dynamics* [1961, 15]). Indeed, Forrester published accounts of modeling a company (1961), a city (1969), and the world (1971) in which the model architecture remains very much the same. This generic approach removes a crucial bottleneck: the model is not intended to represent detailed mechanisms.[32] Hence, knowledge and evidence about such mechanisms are dispensable. Moreover, mastering the mathematics of complex interactions also becomes dispensable because the computer simply grinds out the model behavior. Therefore, system dynamics combines mathematics and technology in a way that creates an effective engine for prediction.

Forrester's world model devises five system levels: population, capital investment, natural resources, fraction of capital devoted to agriculture, and pollution. Because the exponential growth of any subsystem will finally outstrip the resources, the phenomena of overshoot and collapse are typical for system dynamics models. In the case of world dynamics, population growth is assumed to be exponential and will therefore play the dominating role, growing until the quality of life becomes too small given the limited reservoir of nature.[33] Forrester's results show that exponential growth leads to collapse due to limited resources (not very surprising). Depending on the speed of population growth, the model predicts that this will happen within the next fifty to one hundred years. *LtG* works with a refined version of Forrester's world model and specifies fifteen basic and derived level variables. The subcomponents are modeled in more detail, but the main tendencies coincide with Forrester's coarser model. Figure 5.4 displays some typically graphical output, showing phenomena not unlike those of industrial dynamics (figure 5.3).

Behind the generality of system dynamics and the apparent ease of making predictions—that means, once the computer is available and the model is set up—there lurks a premise: scientific predictions are relevant in the context of decision and policy to the extent that the predictions rest on well-established, commonly accepted methods and assumptions.[34] At the same time, *LtG* was part of a new culture of prediction whose selling point was that established methods were inadequate for making predictions. When Forrester, Meadows, and their colleagues promoted system dynamics and *LtG*,

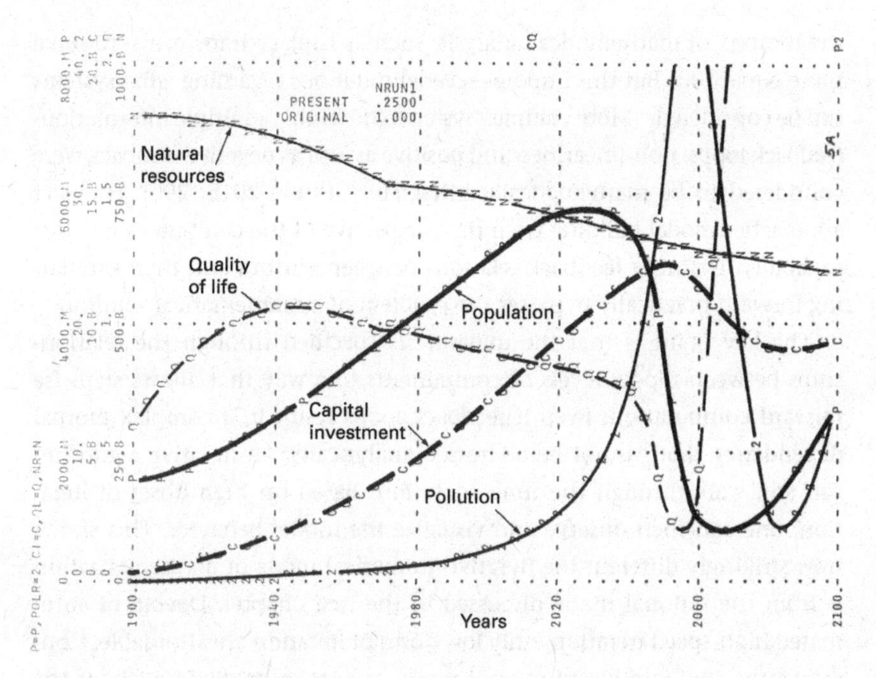

Figure 5.4

The reduced usage rate of natural resources leads to a pollution crisis. Forrester (1971, 75), chapter 4, "Limits to Growth," figure 4.5.

they had to balance these counteracting demands—and perform a tightrope walk. Meadows et al. (1972) highlight two advantages of a formal model:

> First, every assumption we make is written in a precise form so that it is open to inspection and criticism by all. Second, after the assumptions have been scrutinized, discussed, and revised to agree with our best current knowledge, their implications for the future behavior of the world system can be traced without error by a computer, no matter how complicated they become. (22)

In short, prediction requires mathematization—an established feature of science, but making the math model predictive requires working with computer models. Thus, these models are necessary to understand how "the traditions of civilization can be altered to become compatible with global equilibrium" (Forrester 1971, 125).[35] According to this perspective, predictions based on computer models become a *sine qua non* for predicting—and controlling—the future of humankind.

The usual mathematics of prediction and control modeled the dynamics under consideration with the help of differential equations and then used the toolbox of mathematical analysis, such as Laplace transforms, to solve these equations. But this imposes severe limitations regarding what systems can be considered. "More complex systems containing multiple information-feedback loops, non-linearities, and positive as well as negative feedback were considered to be mathematically intractable" (Kline 2018, 290). Forrester approaches model building from the perspective of the computer's iterative capability. Defining feedback relations between entities and then simulating forward practically bypasses the problem of a mathematical solution.

The key point is that the dynamic is specified through the relationships between model levels or components in a way that allows stepwise forward computation. Even if feedback loops result in a complex mutual dependency that cannot be addressed analytically, an iterative algorithm can still walk through the time evolution. Based on high doses of iteration, one can then observe and visualize the model behavior. This shows how strikingly different the iterative–numerical mode of mathematization is from the rational mode discussed in the first chapter. Devoid of automated high-speed iteration, only low doses of iteration are affordable. Consequently, the iterative–numerical mode can make predictions about the behavior of systems far beyond the reach of traditional mathematics. At the same time, this approach is tied to iterations because it loses the generality of mathematical solutions—that is, many properties can be investigated only by first running simulations and then inspecting their outcomes.

To the mathematical community, this approach looked a bit outlandish. It had not grown from mathematics; rather, it came directly from the technology of the computer. How then does Forrester's system dynamics relate to more traditional mathematical approaches?

Richard H. Day, working at the Mathematical Research Center, University of Wisconsin, compared Forrester's system dynamics to conventional mathematical concepts and language (Day 1974). He discussed an example that he could use to translate between the two approaches. Of course, this strategy meant that only a model of low complexity was eligible. Day chose the Solow growth model, well known in economics as a toy model for growth.[36] When he rephrased Solow's model into Forrester language, Day found "with perhaps a touch of irony" that this made the model appear much more complex. The system dynamics model first needed the three entities plus valve-controlled flows between them and then delivered a prediction (conditional on all parameters set) but did not provide any principled insight into the relationship of the variables. On the other hand, according to Day (1974):

> Its not inconsiderable advantage is that every structural hypothesis in a model can be represented graphically. Its disadvantage is that it increases the number of variables and equations, in this way possibly increasing the appearance of complexity. (261)

In the numerical–iterative culture of prediction, this disadvantage is irrelevant or even a rhetorical advantage because complexity now indicates the need to bring in computer methods. The digital computer almost inverts the landscape of tractable algorithms.[37] Digital computers perform feedback loops and iterations automatically and at high speed. Thus, highly iterative algorithms—something mathematicians had previously tried to avoid or restrict—become attractive through the computer. Iteration is what mainframe computers are good at. The pull toward iteration heavily influences computational methods and mathematical modeling. This warrants calling the mainframe culture of prediction an *iterative*–numerical culture.

5.3 Contested Road to Prediction

Indeed, *LtG* attracted a great deal of attention from a wide audience and became a bestseller in Europe, the United States, and Japan. It also triggered a number of follow-up studies from a growing number of modeling groups. Ashley (1983) depicts the dramatic impression the study made:[38]

> Within months the idea of world modeling (once considered thoroughly audacious if considered at all) becomes imaginable to governments and general publics, as well as to professional social scientists. And within a few more months numerous follow-on world modeling studies are underway around the globe. What Karl Deutsch would proclaim "a new stage . . . in the study of world affairs," the "stage of large-scale computer-based world models," is upon us. (496, referencing Deutsch [1977, 1])

However, what looked like one movement from the actor's perspective was actually composed of heterogeneous developments. The blossoming of world models did not last long. After a second generation of models,[39] the entire movement started fading (cf. also Andersson [2018, 184]). In a sense, it became a victim of its own success. Different groups developed models that elaborated on different aspects designed to predict events or trends in line with their own interests. One example is the Bariloche model, named after an Argentinian region and built by a group of Latin American researchers (Herrera et al. 1976). They constructed a more detailed model showing that the impending catastrophe could be avoided with just minimal sacrifices from third-world countries. Another instance is the modeling group led by sociologist Deutsch that argued that extant world models did not take adequate account of societal dynamics (Deutsch et al. 1977, 27). A third and last example is the Club of Rome's second report *Mankind at the Turning Point* (Mesarovic and Pestel 1974). The Club wanted to react to the criticism that the first report was a "model of doom" with an all too predictable outcome. Therefore, it commissioned "a less deterministic, more disaggregated, multilevel world model" (Deutsch et al. 1977, 21). The International Social Science Council (ISSC) started an initiative to integrate and synthesize the growing model zoo (see the entire volume of Deutsch et al. [1977]). They acknowledged the problems—"We have problems of the compatibility of concepts, data, existing models, programs, and computers" (Deutsch et al. 1977, 10)—but could not solve them.

It turns out that computer modeling is a tool with an inbuilt tendency to diversification. Computer modeling is an exercise that is hardly regulated by scientific conventions. It does not require the kind of disciplinary training that rational mechanics required. Furthermore, the building of models requires working under computer-imposed conditions of a technical nature with which active modelers have to comply, especially in mastering software. For instance, Forrester's (1980, chapter 8) group at Sloan had developed the DYNAMO compiler and made it publicly available. However, using it meant following the predesigned paths; not using it created difficulties in accessing the computer—difficulties that researchers without much technical expertise in computers wanted to avoid. And with mainframe computers changing from expensive and scarce to available resources, model sprawl set in—featuring a number of different factors such as more natural resources, more effects of economic growth, or societal dynamics. In short, a plurality of world models evolved, each one predictive but hardly compatible with each other. The result was that any single model looked arbitrary.

The critical perspective on arbitrariness also concerned system dynamics itself. Of course, Forrester was proud of system dynamics' general nature. Others raised doubts. Taken with a grain of salt, how can a model that captures beer supply also have something meaningful to say about the predicament of humankind? Two salient lines of criticism were that available data were insufficient to build a predictive model, and that the prediction was worthless anyway because the behavior was overdetermined from the outset (overshoot and collapse).[40] In particular, if the outcome depends on conditions and assumptions that are not well known, the prediction does not seem trustworthy—with the exception of those behavioral features that remain invariant under changing conditions. This is exactly what Forrester and the *LtG* study highlighted: with exponential population growth, a collapse will happen, quite independently from data about other parts of the dynamics that are not available anyway. From the start, the robustness of this behavior was taken for granted.[41] "Forrester was among the first to insist that computer models could serve important policy purposes even in the absence of good data" (Edwards 2000, 239).[42]

Forrester, who was seasoned in warding off attacks since his Whirlwind days, anticipated both criticisms. He insisted that being predictive was all-important and *required* computer-based world modeling. World modeling had to perform a tightrope walk. There was no alternative way to obtain predictions, but these predictions did not depend on the data and modeling assumptions (except for population growth). The resulting claim sounds almost paradoxical: the world model is a first step that is merely tentative but also without alternative. Forrester devotes considerable space in his study to arguing that his model is far from perfect and that it provides a starting point for later improvements. He claims only a modest status for his model—that is, validity in a relative sense:

> The theory of world structure . . . may seem oversimplified. On the other hand, the model presented here is probably more complete and explicit than the mental models now being used as a basis for world and national planning. The human mind is not adapted to interpreting the behavior of social systems. (Forrester 1971, 123)[43]

Forrester does not just admit that his model can claim only a preliminary status. He also actively calls for improved versions. This move was a clever gambit because all that his modesty took for granted was that his model was on the right track to prediction, even if highly imperfect. From this point of view, the issues of data quality and adequacy of representation were not crucial.

Another line of argument took the opposite standpoint, criticizing Forrester and system dynamics for being too restrictive. Main proponents were experts in management theory, a field with its own established research and institutions. In a sense, this line did not accept Forrester's gambit and felt uneasy with the new culture of prediction and its version of technical rationality. Ansoff and Slevin (1968) illustrate the case; in their reaction to *Industrial Dynamics* (Forrester 1961), they argue that the entire enterprise is misguided. The system dynamical mathematization will not make management more objective because setting up and adapting the model is even more subjective than traditional judgments by managers. In other words, business schools should not transfer management competence to engineers and modelers.

Herbert Simon (1996), the prominent economist, political scientist, and cognitive scientist, attacks the rationality in a more direct way. He remembers when he was part of the President's Scientific Advisory Committee and Forrester presented to this committee on a promotional tour shortly before the Club of Rome study was published.

> My reaction was one of annoyance at this brash engineer who thought he knew how to predict social phenomena. In the discussion, I pointed out a number of the naive features of the Club of Rome model, but the matter ended, more or less, with that. (Simon 1996, 301)

Simon is struck by the fact that Forrester orients his approach toward the iterative capabilities of the instrument and does not pay attention to how social phenomena should be represented. Indeed, Forrester ignored the existing discourse on rationality and formal methods, provoking Simon's remarks on the brash engineer. Simon pioneers a different approach to computer-based modeling, using a model as a symbolic representation of human reasoning (for Simon's work on artificial intelligence and bounded rationality, see Crowther-Heyck and Simon, [2005, chapter 9]). Kline (2018) rightly observes: "In the United States, social scientists in the related field of management science critiqued engineer Jay Forrester for creating a modeling technique at MIT's business school outside the culture of social science" (302–303). Forrester is a showcase member of the mainframe culture of prediction who valued behavioral characteristics and prediction over description and representation. The debate on which grounds computer methods should target predictions echoes the rational–empirical tension explored in chapters 2 and 3 in which the grounds on which mathematical methods arrive at predictions were highly contested. In the mainframe culture, again, a struggle is taking place over who has, and for what reasons has, the authority to define what is "rational."

5.4 Open or Closed Future

Forrester's work and the *LtG* study have served as a sample of the mainframe culture of prediction. This sample reveals typical features of the coevolution of social organization, technology, and the iterative–numerical mode of prediction, thus confirming the findings on computational chemistry in chapter 4. At the same time, this particular sample is very special insofar as it is about the future—not just like any temporal prediction is about what will happen in the future, but in a grander sense. It is about predicting the future of the world. What could be a more ambitious scheme for prediction?

As we saw earlier in the chapter, the future was a hot topic at the time, and the *LtG* can surely be seen as a part of the future movement. Interestingly, this movement reacted in two markedly different ways when *The Limits to Growth* opened the door to scientifically predicting the future. One camp welcomed with open arms the idea of predicting the future with computer models and the game-changing effect of the *LtG* report. This camp was in the mainstream of planning and management that sought ways to enlarge the range of scientific predictions. In the early 1970s, with *LtG* as an influential paradigm, dealing with the future became professionalized and defined increasingly through technologies and methods of computer modeling. In other words, futurology moved under the umbrella of the mainframe culture of prediction.

The second camp was skeptical of prediction. According to this point of view, *LtG*, and computer-model-based prediction in more general, narrow down thinking about the future because any modeling approach conceives of the future as the temporal processing of modeling assumptions. This deprives the future of its proper open character. The second camp included intellectuals as diverse as the urban theorist Lewis Mumford, the journalist Robert Jungk, the Marxist Ossip Flechtheim, the activist-economist-sociologist couple Elise and Kenneth Boulding (the cofounder of general systems theory), and the American economist John McHale. What united them was that, in one way or the other, they called for an alternative notion of the future: "To Flechtheim, the future was not a science of prediction, but a new and more systematic utopian reflection on the present" (Andersson 2012, 1412).[44] Taken with a grain of salt, the future movement was fighting a cultural battle over whether to integrate with or separate from the mainframe computer culture of prediction.

This battle, of course, is over the question whether mathematization (computer modeling) either eliminates the future from scientific discourse or rather makes it relevant. And answers to this question are rooted in basic positions about what thinking about the future is or should be. The opposing standpoints of philosopher Hannah Arendt (1906–1975) and historian Reinhart Koselleck (1923–2006) might exemplify this point.[45] Although both reason about modernity in a very general sense and do not have something like the mainframe culture of prediction in mind, their positions fit this issue quite strikingly.

For Arendt (1951; 1958) and others such as Walter Benjamin, predicting the future poses a fundamental political problem: if the future can be calculated or predicted from the present in a mechanistic (algorithmic) way, so Arendt holds, progress becomes a totalitarian force. Forecasting the weather is often useful. Such predictions, of course, are based on the firm assumption that the driving forces of the weather remain the same. In the very same way, predicting the human future by similar methods means neglecting the inherent possibility of change. The future is then a mere extrapolation (if a complex one) from the present. Arendt diagnoses a pervasive human crisis linked to modernity because humanity has replaced all eschatological and moral notions with the totalizing idea of constant progress.

The German historian Reinhart Koselleck takes the opposite stance. He views this movement not as a crisis but as a liberation. For him, bringing together future and (scientific) prediction does not so much implement a dangerously totalizing idea of progress but a liberation from the grip of (religious) tradition. Koselleck asks what kind of experience is opened up by the emergence of modernity?[46] "For Koselleck, in striking contrast to Hannah Arendt . . . it was the separating out of a secular and manmade future from the grip of Christianity that gave the future its political relevance" (Andersson 2018, 14–15). World modeling, part of the mainframe culture of prediction, belongs to what Koselleck calls the scientification of the future. According to him, this process has replaced one form of power with another and closes a dangerously open future (Koselleck 1981, 176–179).

In summary, our examination of the mainframe culture of prediction leads to the question of whether mathematization, or computerization, threatens to further close a future that should remain open or rather helps to close a future—hence making it predictable—that is dangerously open. This is an eminently political question of undiminished importance.

6 The Fluidity of Computational Models

This chapter pioneers the history and sociology of computational fluid dynamics (CFD) software.[1] CFD emerged in the mainframe culture of prediction and built up, to this day, one of the major applications for computer simulations. Although computing centers were largely unconnected over several decades, and although getting access to a computer presented a major bottleneck up to the 1960s (see chapter 4), there was, nonetheless, a degree of travel activity. This chapter highlights how, during the mainframe era, the movement of models was linked closely to persons and their work, whereas during the era that followed, networked computers came with a different social organization and travel pattern of software. As a piece of scholarly work on computational models, our aim here is to explore new ground and indicates potential directions for further research.[2]

In this chapter, we argue that the commonly used metaphor of circulation overlooks the social epistemology of the use and repurposing of models. Instead, we want to discuss the ways in which models travel, showing that, by traveling, they change to their very core.[3] The metaphor of travel offers more rhetorical space in which to discuss how models change when used by different scientific communities. In addition to offering an account of the effects of movement on different communities, this chapter also lays out the technologies from computer hardware and software to mathematical modeling techniques that allow such movement to occur. A computational model has a medium that can facilitate or interfere (or both in different ways) its adoption and adaptation by new users. The development of networked desktop computers has greatly boosted the travel of CFD models through common and commercial software packages, internet model hubs, and changing disciplinary reward structures. Nonetheless, CFD models and their construction and adaptation still mean different things to different users. This chapter examines this variety of social and epistemic phenomena accompanying the travel of CFD models. We examine the complex interaction of software, technology, and social organization along three different configurations: work at a relatively isolated, military-related supercomputing center (Harlow at Los Alamos), university-based work on building up an international community (Spalding at London and Purdue), and a more recent applied field (traffic simulation) that uses CFD as a tool for prediction.

6.1 The Partly Missing History

A look at a science or engineering textbook will often reveal that it offers a kind of canonical history of the technique or theory being explained. For example, a mechanics textbook will often offer a glimpse of Galileo and Newton as a way of showing that, despite being in the twentieth or twenty-first century, these ideas have been around awhile, have been proven, and have famous, or sometimes should-be-famous, names attached to them. It is common to find thermodynamics referring to the conundrum of the steam engine and a parade of international figures including Sadi Carnot, James Joule, Rudolf Clausius, and Walter Nernst. These origin stories are often whiggish and vastly oversimplified, but they serve a purpose: to establish the staying power and the importance over time of the problems and solutions at hand. Upon opening a textbook in computational fluid mechanics (hereafter referred to as CFD), typically no such origin story is offered. This is curious. To what can we attribute such a lack? Is it the relative youth of the field? This cannot be the case because fluid dynamics in its precomputer form has a long, distinguished, and commonly cited canonical history to draw on (the Bernoullis and such). The story of computation needs only to be a postscript. Perhaps not attaching them to any particular set of problems is part of the effort to show the flexibility and mobility of CFD techniques? But this cannot be true either because the problem sets that follow will inevitably show a range of specific engineering problems that benefit from the application of CFD. This would also imply a particularly historical reflexivity on the part of engineering textbook authors that seems unlikely. CFD is also not a basic engineering science; it is typically taught only to advanced undergraduates or graduate students for whom the socialization and disciplinary inculcation that an origin story provides is viewed as unnecessary. This may be true, and it is common to see origin stories missing from other computational techniques such as finite element textbooks. But it is not the case that CFD has no easy-to-tell or revealing origin story. CFD textbook authors would do well to understand, simplify, and bastardize the origin story of CFD because it does make for good storytelling and inculcates the value of persistence in the face of difficult problem-solving.

However, there are also a number of challenging methodological difficulties. First, it is hard to create a balanced account of CFD software. There is a dark, hardly visible side to it that deserves more appreciation. It is uncontroversial to say that "CFD can be traced to the early attempts to numerically solve the Euler equations in order to predict effects of bomb blast waves following World War II at the beginning of the Cold War" (McDonough 2007, Prologue). These attempts at prediction happened in Los Alamos and targeted the conditions relevant for atomic bomb explosions. Of particular interest as well as difficulty were interactions between hydro- and thermodynamics. All this work was classified, and much of it still is—such as material properties at high temperatures. Only stripped-down versions of code have been published. Naturally, this impedes historical work. How unpublished code travels can hardly be accounted for. As a consequence, the work by Spalding in Huntsville probably gets more credit here than it deserves relative to the work done at Los Alamos. How one should compensate for this, however, is an open question.

Second, comparison can usefully sharpen the case for how code travels. Of course, there are more or less parallel cases such as that of density functional theory in which a large number of software packages are used in a tightly networked infrastructure (see chapter 4 on computational chemistry). Comparison with nonsoftware tools that travel is arguably of equal value. One pertinent point for comparison is David Kaiser's work on "the dispersion of Feynman diagrams" (2005).[4] These diagrams originated from Feynman's work, also at Los Alamos, and spread out like tools to influence education and communication. Kaiser's point is that the diagrams traveled mainly through a network of postdocs. Over the years, different groups of scientists learned to use these diagrams in different ways. Kaiser stresses the differences between his work and that of Bruno Latour:[5] "Whereas Latour emphasizes 'optical consistency' (even 'immutability') as an essential feature of why diagrams and other scientific inscriptions carry so much force among scientists, I focus instead on unfolding variations within their work—on the production and magnification of local differences" (Kaiser 2005, 7). CFD presents a different case because when code travels, this means that it is changing and developing. On the one hand, different codes are predictably inconsistent with each other. On the other hand, all CFD codes have a backbone of mathematical theory, of computer programming techniques, of predictive success, and of predictive failure. Finally, comparison could and should include the different cultures of prediction as discussed in this book. Those rare examples in literature on CFD that mention some history address only the history of mathematics, including that of algorithms, but not that of software.[6] The following text contributes to changing this.[7]

One of the benefits of mathematical models is the ease with which they can be tested and applied to different kinds of problems, even when new applications may have little to do with the physical systems these models were originally designed to represent. In this chapter, we offer a case study in computational fluid dynamics (CFD) to show the ways that CFD models were developed and how they moved to tackle new problems. We argue that by traveling to new users and fields, the models themselves changed on many different levels ranging from the entities they could represent to the kinds of code (both algorithms and programming languages) used to modify and add on to them. CFD is an ideal case study for examining the travel of computational models because its use is so wide-ranging, being applied in everything from mechanical engineering and physics to economics, traffic simulations, and accident reconstruction (e.g., the Hajj stampede of 2015).

6.2 Cold War Research

Many computational techniques have long precomputer histories, and, in the case of CFD, there is a very long and distinguished history in fluid dynamics. As with many canonical histories, fluid dynamics—or, sometimes, fluid mechanics—traces its origins to the Greeks, starting with Archimedes and a theory of buoyancy. Things really get going in the early modern period with the research of Galileo's "disciple" Torricelli, the oft-forgotten Versailles-based Mariotte, and the famous Bernoulli brothers, followed by the mathematical antics of Jean D'Alembert and Leonhard Euler.[8]

Despite much research in the nineteenth century, including the work of Helmholtz, the canonical story of fluid mechanics is really characterized by the Torricelli to D'Alembert period.

CFD really begins in the Cold War and is continuous with research prior to World War II. However, the field of fluid dynamics had expanded significantly in the prewar period due to the invention of the airplane. This focused on the challenges of aerodynamics, with the work of Prandtl as a key. This work was also carried out by his student von Karman, who emigrated to the United States in 1930 to organize the Aeronautical (later Jet Propulsion) Laboratory at Caltech.

The computational story has to begin in the Cold War because it is coterminous with the computer and also provides the basis of problems that needed solving. It begins at a famous site, the Ames Research Center in Sunnyvale, California. In the early 1950s, Ames was the secondary research facility of the National Advisory Committee for Aeronautics (later NASA). In the 1950s, a great deal of Cold War–oriented aerodynamics work was being performed there. One problem in particular defied conventional modeling attempts and would be passed on to more dedicated computing facilities at Los Alamos by the early 1960s.

This problem was as follows. While doing research on the nose of a long-range ballistic missile, researchers found that modeling the pressure distribution was troubling because of a counterintuitive finding: mathematical analysis seemed to suggest that a blunt nose would experience less aerodynamic heating than a sharp slender one.[9] The issue of aerodynamic heating, related directly to pressure, was an important one because the nose cone of the missile needed to remain intact during re-entry through Earth's atmosphere. Temperatures reached 7,000 K, so the structural integrity of the body was at stake in the modeling. The question became how best to model the distribution of pressures and temperatures around that blunt nose. Mathematically the problem is messy. The steady flow near the nose is subsonic, whereas the downstream flow is supersonic. This means that two incoherent sets of equations are needed to model the phenomenon. The subsonic flow is represented by elliptical partial differential equations (PDEs), whereas the supersonic is modeled with hyperbolic PDEs. But these two regions meet at a boundary called the sonic line (see figure 6.1). Therefore, the model needed to be mathematically consistent at this boundary. The effort to model the phenomena was driven by a design question—what

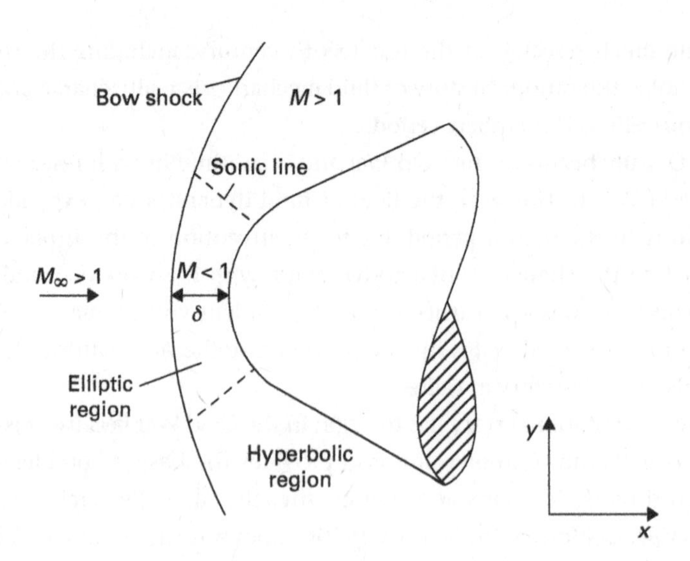

Figure 6.1
Qualitative aspects of flow over a supersonic blunt body (original caption of fig. 1.1, Anderson [2009, 4]). Courtesy: Springer.

is the optimal shape of the nose to have the best, and most predictable (i.e., smoothest), pressure and temperature distributions? Because the models produced at Ames could not be solved by hand, the problem was passed on to the computational facilities at Los Alamos.[10]

At Los Alamos, the problem was delivered to physicist Frank Harlow (1928–2016).[11] Harlow set up a model that could be solved by using a computer. Rather than focus on the hand-off of values from the elliptic model to the hyperbolic, Harlow drew a mesh over the whole cone, and tried to calculate the pressures at the nodes of the mesh. In this way, he traded the (unattainable) general mathematical solution for a stepwise procedure working with many nodes. In other words, he transformed the problem into something that lends itself to an iterative–numerical strategy. This sort of transformation was the centerpiece of the iterative–numerical mode of prediction.[12] The computer could solve these meshes iteratively and avoid the difficulties of insolvable PDEs. Harlow combined the Eulerian approach using a fixed mesh with a Lagrangian approach in which the mesh followed the flow. As a result, the fluid masses moved with the Lagrangian, whereas the mesh itself stayed stationary on the nose cone.

Harlow also focused on producing a graphical output or picture of the pressure distributions. For him, this would be a significant development:

that the best way to understand the output of a fluid model was not as a value table, although that was easy enough to produce and was often needed for design purposes. But, rather, to understand the situation at hand, a visual display would be very helpful. This orientation set in motion a priority for CFD models to produce visual displays of the complex phenomena they were designed to create (and predict). As display and printing technologies have improved, these graphical outputs have become all the more important and sophisticated. We have not been able to find the original output of the nose cone but have found relatively contemporary work by Harlow on another set of problems that he was modeling using CFD in the early 1960s; see figure 6.2.

Harlow's techniques were an important beginning, but Harlow was a physicist inside a national weapons lab. He was neither committed to distributing the methods he developed nor were the projects always easy to declassify. Harlow did define the method in the following way: computational fluid dynamics is, in part, the art of replacing the governing, and unsolvable, partial differential equations of fluid flow with numbers, usually experimentally derived, and advancing these numbers (guided by equations) in time and space to obtain a final numerical description of the complete flow field of interest. Harlow also offered a how-to for building fluid flow models that will be important as CFD travels to new questions.

- Define and represent the system geometry.
- Divide the area or volume into cells (a mesh).
- Define the governing equations of the system (e.g., in the case of the cone, this was the elliptical and hyperbolic equations of subsonic and supersonic flow).
- Set the boundary conditions.
- Run the simulation on the computer.
- Produce a visually informative output.
- Analyze the results.
- Tweak the model's representational features and boundary conditions if needed and rerun the simulation.

This list continues to be descriptive of the process of making CFD models to this day, though there are additional steps once the simulation has been run to produce the most visually informative output, steps that are typically called "postprocessing."

Figure 6.2
Harlow's 1965 *Science* article modeling fluid flow at a sluice gate. Cover of *Science* volume 149, issue 3688. The original caption (p. 1092) is: "Surging water under a sluice gate is simulated by computer calculations. The sequence of figures shows unretouched computer output, illustrating early stages in the development of a hydraulic jump from a backward-breaking wave. Such numerical solutions of the full nonlinear, time-dependent Navier-Stokes equations make possible the detailed study of this and numerous other problems in fluid dynamics." Reprinted with permission from AAAS.

6.3 Imperial College as a Different Kind of Site for Knowledge Production

Even after Harlow's retirement, the CFD work at Los Alamos was little known outside the US Department of Defense. However, other researchers were working on CFD. This section focuses on a very prolific group at Imperial College (IC) in London: a group who produced a large number of CFD disciplines in the 1970s and 1980s. It was led by Brian Spalding (1923–2016) from New Malden, a London suburb. Spalding had a 1952 PhD from Cambridge that had been supported by an Imperial Chemical Industries (ICI) scholarship. Spalding wrote his dissertation on a method of unifying von Karman's hydrodynamics, Georgii Kruzhilin's heat transfer equations, and Ernest Eckert's mass transfer models to produce a general mathematical theory of heat and mass transfer with and without combustion. The insight gained by Spalding's thesis was that chemical reaction rate constants do not influence combustion until a critical rate of mass transfer is reached. In 1954, he was hired as a reader in applied heat at Imperial College, and by 1958, he was a professor of heat transfer. Throughout the 1960s, he built up a large research group (averaging around thirty members) with students from all over the world, several of whom followed in his footsteps with ICI fellowships. Over his career, Spalding trained approximately one hundred graduate students (i.e., around three new PhDs per year).

Here are three examples:

- In 1964, Suhas Patankar arrived from India on an Imperial College scholarship and put Spalding's theory into FORTRAN code.
- In 1965, Akshai Runchal also arrived on an Imperial College scholarship. Runchal had initially planned to work on the drying of sprays, thinking that was something of interest to ICI (which is largely a paint company); Spalding told him that it was not interesting but that he should come and they would find him a project. He worked on finite difference solutions for low Reynolds number fluids.
- In 1965, Micha Wolfshtein came from Israel to work on High Reynolds number flows.

Runchal was working on literature reviews and decided that the math department might be able to help him. There, he learned about the promise of using finite-difference methods to approximate the Navier–Stokes equations. Runchal and Wolfshtein discovered that they were both working on finite difference solutions—Runchal on low Reynolds number systems (i.e., laminar flow) and Wolfshtein on high Reynolds numbers (turbulent flow). Spalding talked to them both and got them to think about the problem physically—the nodes in the finite difference grid represented tanks that fluxed with other tanks by way of tubes. Physical intuitions were the proper way to think, according to Spalding.

Runchal and Wolfshtein finished their theses within a year of each other—theses on similar problems but with different solutions. Runchal then went on to teach at the Indian Institute of Technology (IIT) Kanpur (although he would return to Imperial College two to three years later), whereas Wolfshtein went to the Technion. These young engineering professors spread the CFD word very quickly.

Spalding decided in the late 1960s that he needed to work harder on getting the tools that he and his students were developing into working engineers' hands. So, he created a "post-experience course" of continuing education for industrial engineers to keep them up to date with the methods he had been developing since 1965.[13] Spalding, Runchal, and another student, David Gosman, then wrote a textbook (Gosman et al. 1969) to further inform potential users about the new developments in CFD. The book contained the text of the ANSWER code in FORTRAN for users to program their local machines. This is significantly different from having a software package that travels to the user. Training in software implementation and modification was a necessary activity on the side of the users that the makers of software first had to invent.

In 1969, Spalding and several of his students created a consulting company to do both CFD contract work for industry and to help train industrial engineers in the methods. The firm was called Combustion, Heat and Mass Transfer, Ltd., or CHAM. In the early 1970s, Spalding turned his attention to 3-D turbulence and heat transfer in chemical reactions. He started with the well-known work by Kolmogorov and Prandtl (in the original Russian and German) and used experimental data to derive constants that could quantize intractable equations. This allowed him to produce a new group of algorithms that were computationally efficient and made accurate predictions. These algorithms were written primarily by Patankar and were called SIMPLE. Even today, nearly all commercial CFD packages use them. As the CFD codes CHAM used grew and their clientele expanded, CHAM became too big for its offices at IC and moved to a storefront in New Malden in 1974. Patankar and Runchal, who had returned from India, moved to full-time work at the CHAM offices, where they developed more tools. Gosman, Laudner, Whitelaw, and Lockwood (all Spalding PhDs) also worked in the New Malden office. Two more CFD firms were spun off Spalding's code: David Tatnall and Harvey Rosten founded Flomerics; and using Spalding's code, Jim Swithenbank started the firm Creare LLC that produced FLUENT, one of the most commonly used CFD packages today.

Unlike Harlow, who locked CFD inside a Cold War institution, Spalding purposefully built up a CFD community that facilitated the movement of CFD to new types of problems. Spalding did this by focusing on publications ranging from textbooks to new journals that would make the methods accessible to anyone able to get their hands on and read the English-language materials. Spalding also engaged the growing interest among engineering professionals in continuing education by offering "post-experience" courses that promised to keep engineers who were already working up to date on new methods. He cultivated a global (albeit English-language) community that brought university researchers and faculty to London for instruction and experience as well as sending out representatives to teach courses at diverse institutions, including Göttingen, Penn State, Stanford, Berkeley, Brown, Caltech, Tsinhua, IIT Kanpur, the Technion, and even Los Alamos itself. The movement of CFD had come full circle from Spalding back to Harlow's home, where Tony Hirt had taken over for Harlow and was working to declassify and distribute CFD code to make it available to researchers outside the Department of Defense system. Through all these transitions, CFD capacity increased and was used to solve more kinds of problems. Through travel, it was transformed.

One methodological tool we have been working on is the production of social network analysis maps in order to understand the CFD community at various times in its development. When making a map of the social network that Spalding created, it is easy to see that his role was central and that the network was well connected and large. We used the program GEPHI to create this map that shows researchers' closeness, influence, and centrality. We did not use co-citation analysis. Figure 6.3 shows the social network of the CFD community. One can read off important information, even without taking into account the exact location of the nodes, which is generally not well-defined (i.e., might vary depending on the exact specification of the

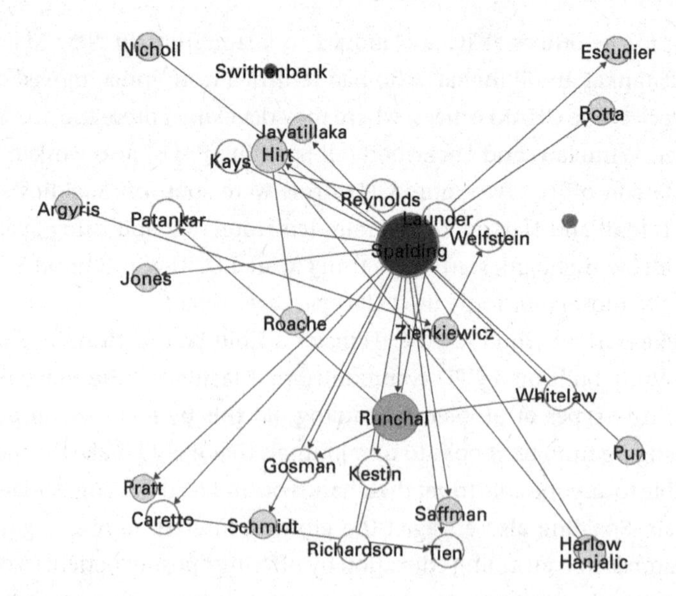

Figure 6.3
A social network map of the CFD community. Courtesy: authors.

mapping algorithm). Clearly, Spalding forms the central node. Although there are more links than just to and from Spalding, the shortest connection between two nodes often runs through him, stressing the influence of his position. Overall, the nodes are well connected, with only a few that are unconnected. Thus, it appears to be a single community of researchers. Furthermore, institutional links are not readily apparent, probably superseded by the (joint) traveling of code and people.

When further investigating the movement of code that Spalding facilitated, it is important to see that technical commitments matter for the mobility and flexibility of the methods. Spalding and his students worked through physical analogies, best seen in the interactions between Spalding and Runchal and Wolfshtein. He also modeled simpler, but more generalizable, problems than Harlow had done. His problems were typically steady, rather than transient, flow. They were single- rather than multiple-phase, and the boundary conditions were fixed rather than moveable. As a result, Spalding was designing CFD for the accessible business computers that were commonly available by the 1970s. Harlow had designed for high-end scientific mainframe computers that might be accessible to large engineering firms such as Boeing or Raytheon but did not offer the great possibilities of putting CFD in smaller firms' repertoires of engineering tools. Spalding also attended to distributing the code and eventually making it commercially viable in both proprietary and open code versions. He also set up a consulting business that served to encourage firms to introduce the necessary technology and expertise to perform CFD in-house.

6.4 Moving CFD out into Traffic

To choose just one application of CFD that Spalding and Harlow might be surprised about, we shall look at traffic flow. Predicting traffic behavior is a socially relevant and economically important problem. However, it is governed by many hard-to-model factors. Simulating traffic requires attention to several levels: the "submicroscopic" on which the physical details of each car matter (e.g., braking distances or traction control algorithms); the microscopic on which driver behavior matters (Are drivers driving very close together? Are their decisions impaired by alcohol or distractions such as cell phones?); and the mesoscopic on which users can be treated probabilistically but some system components and geometries need to be attended to. Finally, can the movement of traffic be treated simply as a flow without attending to its constituent parts? Does the last option produce usable and accurate predictions of actual traffic on real roads?

It might be useful to review Harlow's how-to list and see where the differences between actual fluids and traffic as a flow are apparent. Which of the following convert easily to traffic?

- Define and represent the system geometry.
- Divide the area or volume into cells (a mesh).
- Define the governing equations of the system (e.g., in the case of the cone, this was the elliptical and hyperbolic equations of subsonic and supersonic flow).
- Set the boundary conditions.
- Run the simulation on the computer.
- Produce a visually informative output.
- Analyze the results.
- Tweak the model's representational features and boundary conditions if needed and rerun the simulation.

It seems that the one area that might pose challenges is "Define the governing equations of the system" because these are unknown for traffic. But can traffic simply be taken as fluid-like and described mathematically using equations taken from real fluids? This has been the approach of traffic modelers using CFD. There has been no grand justification of the similarity between traffic and fluid—the empirical (even appearance) similarity has been sufficient to try to model traffic with CFD and see whether the models can reproduce the phenomena, especially of traffic shocks (most commonly known as rubbernecking delays in the United States). The justification for working with these models comes mainly from the fact that they produce predictions.

But traffic simulation occupies a different kind of community than CFD developers because its modelers are fundamentally agnostic with regard to their modeling techniques, and they use many different simulation tools to model traffic, often combining different modeling techniques in the same simulation. Figure 6.4 maps the social network of the CFD traffic community. In contrast to the earlier CFD community (figure 6.3), this community is much less well-organized and connected. Figure 6.4 displays a much sparser network of a much newer community with no central node(s), subcommunities, or institutional links. It is unclear how influence flows. One could even discuss whether it is too small to be called a network—that is, whether it is admissible to display only those traffic modelers who work with CFD. We simply acknowledge that the social network analysis gives very tentative results.

Traffic simulation is also highly multidisciplinary, whereas CFD since the 1960s has been located largely within the discipline and departments of mechanical engineering. Traffic simulations do not have a clear departmental or disciplinary home and have required the collaboration of civil and mechanical engineers, computer scientists, applied mathematicians, and even psychologists. Part of the challenge is that simulations need to run very quickly and in real time so as to produce a prediction of traffic conditions that is useful to drivers. Like the weather, no one cares where yesterday's traffic jams and tie-ups were. Drivers also tend to want to know how bad tie-ups will be in order to calculate whether it is worth taking a less direct but alternative route. This means that jams need to be predicted, but so do their longer spatial and temporal consequences. As a result, optimizing the codes for the simulation is more important for this application than

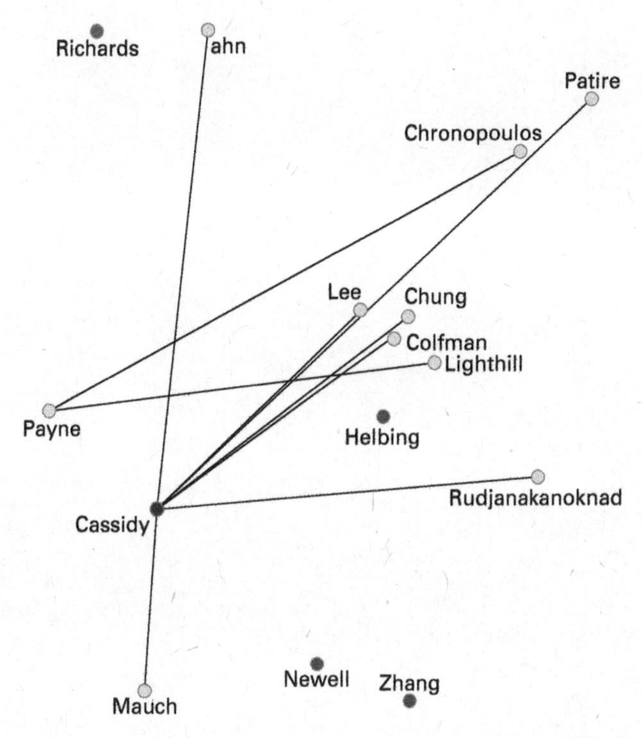

Figure 6.4

A traffic simulation community social network analysis map. Courtesy: authors.

it might be for design processes. This is another way that CFD changes as it travels to new applications.

Since the 1990s, CFD has been a standard package software that can be run on desktop and laptop PCs. This has standardized the kinds of things CFD can do while also opening up a large number of new application fields. Academic researchers in CFD are still rewarded for coming up with new code both to address problems more efficiently and to produce more accurate models. But users of CFD have no incentive to modify code, and in the case of proprietary software, have no access anyway. What is useful to consider is the extent to which CFD has traveled—from clear heat transfer problems to fluid flow to traffic, even to modeling "shockwaves" in the economy—and how this movement has transformed CFD into a general-purpose tool whose predictive powers have superseded questions after representation.

IV The Exploratory–Iterative Mode

7 A Transformation of Bayesian Statistics

If statistics is viewed as a branch of mathematics, it has to be seen as a special branch distinguished by the ways in which it is linked to both societal practices and philosophical positions.[1] Bayesian statistics is exemplary on both counts. Philosophers have discussed Bayesian statistics vigorously and elaborated Bayesianism as a *philosophical* position.[2] Most significantly, Bayesian epistemology analyzes how one should deal with new data in a rational way—that is, the Bayesian standpoint lays claim to capture scientific rationality. Put simply, Bayes's rule[3] is taken as a (normative) principle that prescribes how one should update prior beliefs in light of new evidence. The use of Bayesian approaches in scientific *practice* shows a remarkable career. Despite their philosophical prominence, they remained very much a minority approach in science—but only up to the 1990s when Bayesian methods quickly acquired a high level of popularity in the sciences as well.

We shall describe and analyze this turn. We argue that the success of Bayesian approaches hinges on computational methods that make a class of models predictive that would otherwise lack practical relevance. Philosophically, however, this orientation toward prediction comes at a price. The new computational approaches change Bayesian rationality in an important way: namely, they undercut the interpretation of priors, turning them from an expression of beliefs held prior to new evidence into an adjustable parameter that can be manipulated flexibly by computational machinery—a lubricant for exploratory iteration. Thus, in the case of Bayes, one can see a coevolution of computing technology, an exploratory-iterative mode of prediction, and the conception of rationality.

Section 7.1 briefly introduces the rift in the philosophy and practice of statistics in which the Bayesian and the classical accounts were used, elaborated, and defended by different fields and disciplines—until the popularity of Bayesian methods unfolded in the 1990s. According to the prevailing stance in philosophy, the advantages in terms of rationality account for the upswing of Bayesian methods (section 7.2). Contrary to this view, we claim that it was the move to an iterative-exploratory mode of prediction—on the technological basis of cheap and easily available computers—that drove this upswing. We support this claim with an analysis of the pivotal roles played by Markov chain Monte Carlo methods (section 7.3) together with software packages (section 7.4). In the concluding section 7.5, we present an outlook on the pragmatic stance that has gained ground in the philosophy of statistics over the last two decades.

7.1 Bayes's Popularity

Before appreciating its growing popularity, we shall briefly and, as specialists will rightly bemoan, superficially describe the situation before it started. Bayes's rule captures how to calculate with conditional probabilities. Let $\pi(H)$ stand for the probability of a statement or hypothesis H, $\pi(H \mid D)$ for the conditional probability of H given D. Now, both H and D happen if (for the moment, think of a temporal order) D happens and then H happens given D, or equivalently, H happens and then D happens given H. In other signs: $\pi(D) \cdot \pi(H \mid D) = \pi(H) \cdot \pi(D \mid H)$. Separating $\pi(H \mid D)$ on the left side gives Bayes's rule:

$$\pi(H \mid D) = \pi(H) \cdot \pi(D \mid H) / \pi(D). \qquad (*)$$

It is named after Reverend Thomas Bayes (c. 1701–1761), a Presbyterian minister, philosopher, and statistician. Bayesianism starts out with a special interpretation of this rule. Consider that you have some hypothesis H—for example, that it will rain tomorrow. You do not know for sure, so (in a Bayesian mood) the degree of your belief can be expressed as a probability, $\pi(H)$. Now there arrives new evidence D—say, you stand up next morning and have a look at the sky. This should give you additional evidence and will change your (subjective) probability of rain on this day. Therefore, $\pi(H)$ is also called the "prior" that will be updated. The updated probability, written $\pi_{D(H)}$, of your hypothesis given the data is also called the "posterior." Which numerical value does it have? Bayesians take the position that updating needs to happen by conditionalization. The posterior is the conditional probability: $\pi_{D(H)} = \pi(H \mid D)$. In other words, equation (*) answers the question: the posterior is proportional to the (subjective) prior $\pi(H)$ and to $\pi(D \mid H)$, the so-called likelihood—that is, the probability of the data given your hypothesis (how likely the sky looks like it does in the morning given that it will rain). The term $\pi(D)$ plays the role of a (normalizing) constant.

Bayesianism adopts (*), or a sophisticated variant of it, as a *principle* that *should* guide inferences. In the entry to the *Stanford Encyclopedia*, to present a generic point of view,[4] W. Talbott (2016) identifies the main features of Bayesian epistemology as the introduction of a formal apparatus for inductive logic that uses the laws of probability as coherence constraints on rational degrees of belief. In particular, it takes Bayes's rule (a basic rule for conditional probabilities) as a norm for probabilistic inference, as a *principle of conditionalization*. "What unifies Bayesian epistemology is a conviction that conditionalizing . . . is rationally required in some important contexts—that is, that some sort of conditionalization principle is an important principle governing rational changes in degrees of belief." Famous arguments from Bayesian epistemology, such as the Dutch book argument, set out to show that following Bayes's principle is following a demand of rationality.[5]

The classical camp of, among others, Fisher, Neyman, and Pearson—despite internal differences[6]—criticized mainly two points: first, Bayesian estimations hinge on subjective priors and are therefore not robust. Any robust results would have to take into account the variability of priors—that is, other probability measures that do not correspond to the actual beliefs.[7] Statistical inference should be geared toward the properties of the estimation (such as robustness) rather than rationality according to a system of beliefs. Second, the Bayesian assumptions create high obstacles for practice. Calculating with (*) does not just require the specification of all probabilities involved: the probability of a hypothesis $\pi(H)$, the probability of the data $\pi(D)$ (often expressed via conditioning on different possibilities), and the conditional probability $\pi(D \mid H)$. Crucially, their numerical values have to be computed. In a technical sense, the calculation of posterior probabilities requires an evaluation of very difficult integrals. It is preferable, according to the classical camp, to avoid specification and computation of this kind. Classical statistical modeling aimed to do without priors—such as the famous "null hypothesis" in significance tests that allows researchers to be agnostic.[8]

The rift between the Bayesian and classical camps was reflected by a divide of disciplines. While economics and—of course—philosophy have been (and still are) dominated by Bayesianism, in most natural sciences, it is classical methods that have a stronger footing, although this is certainly not a clean divide. This brought philosophy of science into an odd position. Philosophers worked out a Bayesian normative account, whereas large parts of the sciences apparently did not care but rather continued to prefer classical approaches. Here is a typical opinion from a Bayesian statistician reasoning about why the uptake in scientific practice was so slow.

> Bayesians were still a small and beleaguered band of a hundred or more in the early 1980s. Computations took forever, so most researchers were still limited to "toy" problems and trivialities. Models were not complex enough. The title of a meeting held in 1982, "Practical Bayesian Statistics," was a laughable oxymoron. One of Lindley's students, A. Philip Dawid of University College London, organized the session but admitted that "Bayesian computation of any complexity was still essentially impossible . . . Whatever its philosophical credentials, a common and valid criticism of Bayesianism in those days was its sheer impracticability." (McGrayne [2011, 213–214], quote from interview)[9]

However, the divide changed in a remarkably swift way. Figure 7.1 presents some bibliometric evidence. The data are from the *Web of Science* and count papers appearing in one of five major statistics journals: *The Journal of the Royal Statistical Society B*, *Annals of Statistics*, *Journal of the American Statistical Society*, *Biometrika*, and *Biometrics*. Each point shows the percentage of papers (in one particular year) whose topic contains "Bayes." All five journals come from the classical side that dominated mathematical statistics.

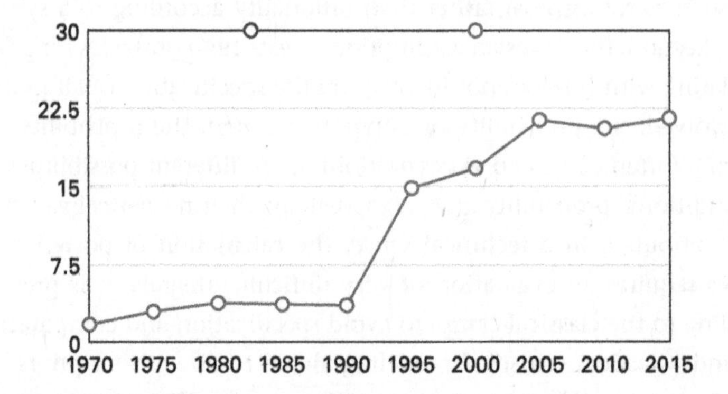

Figure 7.1

The axes are the year of publication and percentage of papers related to Bayes. Ten data points are displayed, one every fifth year, 1970 to 2015. Straight lines are added connecting these points. Courtesy: authors.

The data confirm the outsider role of Bayesian methods in professional scientific statistics, with a consistent share of only 2–4 percent up to the early 1990s. Then, however, there is a rapid rise to a level of about 20 percent.[10]

This picture is conservative because it displays (at least three) very traditional journals that will surely not overrepresent papers inclined toward Bayes. Newly established journals tend to show an even higher share but cannot provide reference points for the mid-twentieth century.[11]

Bayesian methods eked out an existence as a small minority group in the sciences and their statistical approaches—up to the early 1990s. After that, Bayesian methods developed quickly, indeed almost leapt up to become an intensely researched and widely used approach. Since then, the extent of literature on Bayesian methods in the sciences has grown rapidly. One can track this in many forms from journal papers and discussion statements to books and encyclopedia articles. The (Bayesian-inclined) statisticians Bradley Carlin and Thomas Louis (2000), for instance, note:

> An impressive expansion in the number of Bayesian journal articles, conference presentations, courses, seminars, and software packages has occurred in the four years since 1996 . . . Perhaps more importantly, Bayesian methods now find application in a large and expanding number of areas where just a short time ago their implementation and routine use would have seemed a laudable yet absurdly unrealistic goal. (xiii)

Thus, we take it for granted that the uptake and use of Bayesian methods experienced a turn in the 1990s, and we devote the remaining sections to analyzing this turn.

7.2 Rationality or Computation?

The turn did not go unnoticed from the side of either Bayesian statisticians or philosophers. We shall complement the Carlin and Louis quote in the previous section with one taken from the philosophical side. In their volume on the foundations of Bayesianism, Corfield and Williamson (2001) offer an outlook on the field: "Bayesianism has emerged from being thought of as a somewhat radical methodology—for enthusiasts rather than for research scientists—into a widely applied, practical discipline well-integrated into many of the sciences" (3). Scientists and philosophers agree unanimously that the turn happened. The next question is: Why did it happen?

A common viewpoint holds that the main reason for the turn is the rationality of Bayesianism itself that finally became operational thanks to computational methods. Computer-based methods rendered feasible the integrations (e.g., when calculating conditional probabilities in complex models) that Bayes's rule requires; and, as a result, the rule's rationality gained traction.[12] As Corfield and Williamson (2001) put it (looking back on the 1990s), it is only recently that "computers have become powerful enough, and the algorithms efficient enough, to perform the integrations" (4). Although this explanation is plausible, it is crucially incomplete and can therefore easily mislead.

The point is that simply using computational tools does not lead straightforwardly to obtaining those results that had been too difficult to achieve before. The tools are not strictly neutral. Which mathematical tools are used and how they are used might influence the modeling process. This is exactly our key point. Our analysis shows that, thanks to new computational tools, Bayesian methods changed into a new, exploratory–iterative mode of prediction—the mode that readers of chapter 4 have encountered already. Furthermore, we argue, this mode of prediction affects the very nature of Bayesian rationality.

We are well aware that this claim is not easy to substantiate. It ascribes a significance to computational methods that has not always been apparent to practitioners. Early appraisals of the computer and its powers typically took it for granted that the machine would simply carry out logical or arithmetical operations and would not require any new perspective on how mathematical tools lead to predictions. For instance, the statistician Dennis Lindley, a leading advocate of Bayesianism over almost the entire second half of the twentieth century, had seen Bayes's rule as an arithmetic recipe for producing inferences. He considered this procedure to be almost mechanical, given that the integrations could be made feasible (Lindley 1965). Lindley did not see any particular interest in devising computational methods. The next generation of Bayesian-minded statisticians, however, saw things differently. The statistician A. F. M. Smith, a leading voice, argued in a sort of manifesto that it was efficient numerical integration procedures that led to the more widespread use of Bayesian methods (Smith 1984).

Even granted the importance of numerical procedures, it was hard to anticipate just how such procedures would change the method. Identifying the computational tools on which the Bayesian boom is built is straightforward. There is ample evidence in which statisticians write about what created the difference in the 1990s. In fact, there is a remarkable consensus on this point: it was the Markov chain Monte Carlo (MCMC) method that made the difference—and Smith himself provided a key paper.

> When Smith spoke at a workshop in Quebec in June 1989, he showed that Markov chain Monte Carlo could be applied to almost any statistical problem. It was a revelation. Bayesians went into "shock induced by the sheer breadth of the method." By replacing integration with Markov chains, they could finally, after 250 years, calculate realistic priors and likelihood functions and do the difficult calculations needed to get posterior probabilities. (McGrayne 2011, 221–222)[13]

Thus, MCMC opened the door for Bayes to become practically relevant. There is agreement on this point. Now is where the difficult part begins. We shall analyze how MCMC affects the very rationale of Bayesianism.

7.3 The Markov Chain Monte Carlo Revolution: Iteration and Exploration

This section bears the main thrust of the argument. Using MCMC methods, we argue, yields predictions in an iterative–exploratory mode and thus affects the rationale of Bayesian methods.

Integration and Convergence

We start with a brief summary of what MCMC is about. Although this part will be about integration in a technical mathematical sense, we shall keep it on a nonformal level. From the outset, MCMC combines Monte Carlo, which stands for iteratively sounding out mathematical terms, with Markov chains, a class of random processes.

Monte Carlo strategies are based on the law of large numbers. This law states that the expected value of a random variable is approximated by the average of many random trials (that each follow the same probability distribution). In a casino such as the one in Monte Carlo, many people gather around a roulette table after a small series of identical outcomes happens—say three times "13." Some think that the next trial is likely to be the number 13 again; others hold that a different number has to come now. Nonetheless, to the extent that the owner of the casino lets the roulette wheel operate as a (near perfect) generator of random outcomes, the law of large numbers will defy any superstitious beliefs and apply relentlessly: in the long run, the number 13 will make up 1/37 of all numbers (0–36).

It is the simplicity of the example that makes Monte Carlo look trivial. Think of another example: a friend gives you a map of Norway—a country with a famously fractal-like coastline[14]—and asks how large is the area of Norway. In mathematical theory, integration provides the answer; in practice, however, integration can be carried out only for a narrow (and relatively simple) class of functional descriptions. Monte Carlo can help out—just hang the map on the wall, paint a big square around it, and throw darts at the wall. The number of darts that hit the map relative to the number of darts that hit the square approximates the area of the board relative to the square on the wall, which is easy to measure—just count. Of course, this finding hinges on two conditions,[15] namely, that many darts are thrown and that they are distributed randomly across the entire square. For humans, it is hard to fulfill these conditions. But they are almost tailor-made for a computer. One can readily simulate the random procedure and iterate it millions of times—and thereby approximate the integral.[16] Although Monte Carlo is almost tailor-made for the computer insofar as it transforms a problem of *integration* (an operation of calculus) into a problem of *iteration*, it is not immune to the curse of complexity. The simulated value—that is, the fraction of hits among all trials, converges only slowly toward the (unknown) integral. Even after a large number of iterations, the simulated value might not be very accurate, so, despite the high speed of modern computers, Monte Carlo is, in many instances, ineffective.

This is where the second component of MCMC comes into play. Markov chains are processes that move in a space according to rules of a certain type. For every state or location[17] in this space, there is a list of what the possible locations are that the process can reach with the next step, plus a probability distribution according to which the next-step location from this list will be chosen. In other words, the next step of the process depends only on its present location (and the random choice to be made in this step) and not on the history of the process. One can imagine that at each location, the rule for where to possibly move is written on a signpost—such as "with probability x go one step north, with probability 1-x jump hundred steps south"—also called the transition probability for each location. The rules might be complicated, but they never refer to where you come from (never state something like "If you are here for the third time, do . . ."). In other words, moving according to these rules does not require memory: just execute the rules written on the signpost where you stand. Markov chains are often called random processes without memory.

The basic theorem about Markov chains states that such a chain will converge to its stationary (equilibrium) distribution no matter where it started.[18] In other words, in the long run, the process will visit each location in a certain fraction of all steps. Some locations are visited more often and others only rarely—reflecting the equilibrium distribution. The astonishing and crucial observation is that this convergence happens very quickly. MCMC is based on this observation. The pieces come together for numerical integration in the following way: first, there is an unknown integral one can describe but not evaluate, such as the posteriori probabilities in Bayes's rule. Assume one can refashion this integral as the stationary distribution of a Markov chain. Then the recipe is straightforward: simulate the Markov chain for many steps (easy iteration for computers) until it is in equilibrium, and record its value. Reiterate this many times (Monte Carlo). The average overall values obtained then approximate the (unknown) integral. The trick depends on two conditions: first, one must find a way to interpret an unknown integral as an (unknown) equilibrium distribution of a Markov chain. Second, the Markov chain must have reached its stationary distribution before one samples its value. The first condition sounds more difficult than it actually is, whereas the second condition sounds easy but is not. We shall discuss both conditions in turn.

Application: The MCMC Trick

The MCMC method was invented early on in the pioneering times following the creation of the digital computer. It goes back to the work of Nicholas Metropolis, Stanislaw Ulam, and others at Los Alamos and received a classical generalization by Hastings (1970). However, it took another twenty years before MCMC started to take off when examples became available that showed how powerful and flexible the method is—in particular, how doable it is to refashion complicated integrals as stationary distributions of Markov chains. One famous instance is the Ising model that describes how spins (up or down) on a grid interact with their neighbors. The model is famous not only because the simple interaction can lead to phase transitions and other surprising behavior but also because the problem of determining its equilibrium proved to be utterly unsolvable by analytical means and had become a mathematical monument of intractability. It turned out that MCMC could approximate this distribution with a surprisingly moderate effort in modeling as well as computation.[19]

The Ising model is not a singular case. Mathematicians and statisticians quickly realized that the wide applicability of MCMC to long-standing problems of integration changed the game regarding computational tractability. Restrictions to the mathematically convenient could be lowered substantially, and "from now on, we can compare our data with the model that we actually want to use . . . This is surely a revolution" (Clifford 1993, 53). Many actors agree with seeing this as a revolution. Diaconis (2009), for instance, provides an insightful treatise on "The Markov Chain Monte Carlo Revolution."[20] Part of his treatise is worries about the speed of convergence (our second condition) that we shall discuss next. Put plainly, the revolution consisted in how far the limits of mathematically—and statistically—tractable models have been extended.

On the side of the practitioner, the main benefit is flexibility in modeling. Bayes's rule became practical for a wide array of models. Although it required the evaluation of posteriors, thanks to MCMC, they lost their horror. A wide array of Bayesian applications followed the availability of MCMC; computational approaches in fields such as statistical physics, molecular simulation, bioinformatics, or dynamic system analysis started to flourish. Statistician Jeff Gill (2008) called the combination of Bayesianism and MCMC "arguably the most powerful mechanism ever created for processing data and knowledge" (332).[21]

One prominent example from the growing set of MCMC variants is the "Gibbs sampler" for treating inferences involving images with many pixels.[22] It was invented by the Geman brothers as a variant of Monte Carlo and gained enormous traction when Gelfand turned it into an MCMC method.[23]

> The trick was to look at simple distributions one at a time but never look at the whole. The value of each one depended only on the preceding value [The Markov "no memory" property, jl]. Break the problem into tiny pieces that are easy to solve and then do millions of iterations. So, you replace one high-dimensional draw with lots of low-dimensional draws that are easy. The technology was already in place. That's how you break the curse of high-dimensionality. (Gelfand, quote from interview, McGrayne [2011, 221])

Thus, the Gibbs sampler construes a Markov process moving through simple distributions. Thanks to his inventive imagination, Gelfand saw how an intractable object (a high-dimensional distribution) arises from much simpler objects (a process moving through simple distributions). Much like the equilibrium distribution of the Ising model is built from a process that moves through simple distributions (simple flips of one spin), the MCMC trick replaces a computationally intractable object with very many iterations of simpler, tractable objects.[24]

Exploration and Flexibility

MCMC has an iterative nature. It also has an exploratory nature. When proponents such as Smith and Roberts (1993) state that MCMC methods are for "exploring and summarizing posterior distributions in Bayesian statistics" (3), the point about exploration is important. Exploration plays a role on two different levels. First, modeling approaches explore quite generally, and this applies to Bayesian statistics in particular: you always explore what the data are telling you relative to a model that you confront with those data.[25] Exploration of this sort is at the heart of modeling—given the lack of complete knowledge, one explores with the help of models. However, exploration also happens on a second level, exploring the mathematical model itself. And this is where MCMC becomes relevant.

MCMC methods *simulate* relevant properties of mathematical objects (such as integrals or distributions) in numerous iterated trials in order to gain a picture or approximation of these properties. One can compare MCMC with sounding out unknown territory by taking simulated random walks. This modeling approach thus explores the behavior of a (complex) mathematical object such as a posterior distribution with the help of the MCMC machinery. In a way, MCMC explores mathematical properties with the help of probabilistic and iterative means. One can see a *frequentist* element sneaking in here.

However, we want to make an additional point: the speed of MCMC is also an invitation to engage in an exploratory mode of modeling in the following sense. Modelers can work with incompletely specified models that contain parameters that are adjusted only in a feedback loop in which model behavior is observed and modified. Researchers do not need to determine parameters from the beginning; rather, they can adapt them during the process to obtain a better match.[26] For Bayesian modeling, MCMC made exploration on this level feasible. With the help of adjustable parameters, a model can be specified in flexible ways. The MCMC trick brings this flexibility to Bayesian modeling.

A short remark on the timeline. Typically, computational modeling of this explorative sort will be done when computational capacity is easy and cheap to access—including software packages (see section 7.5 of this chapter). On expensive mainframe machines, researchers tend to run only their best models with their best guesses. This accessibility condition started to be fulfilled in many labs and offices from the early 1990s onward, and this coincides with the timeline displayed in figure 7.1.

However, the exploratory–iterative mode affects the Bayesian rationale. The core of Bayesian epistemology, indeed the defining feature for many philosophers, is the subjective stance. The modeling process starts out with one's degrees of belief. We have seen, however, that this characteristic of Bayesian epistemology is fading away over the course of the development of MCMC approaches. Priors now appear as part of the adaptation machinery.[27] Importantly, these parameters lose their interpretation as prior knowledge. To the extent that they are treated as adjustable parameters, the resulting values no longer express (degrees of) *prior* belief but rather correspond to an overall fit of model and data *resulting* from the exploratory–iterative process of modeling. In a nutshell, *the priors cease to be prior.*

We have presented the argument over how using MCMC as a tool undermines the perceived rationality of Bayesianism. We conclude this section by backing up this argument with a second line of thought that supports the claim about the exploratory–iterative nature of MCMC. Now is the time to recall the earlier promise and address the second condition for MCMC: the one that looks innocent but is not, namely, that the Markov chain has reached equilibrium. The results MCMC provides take for granted that the Markov chain has reached stationarity before sampling. If the chain runs one million steps, is that enough? Or, if it is not quite in equilibrium, how does that play out in terms of error bars? Answering these questions is arguably the most important and intricate problem in the validation of MCMC results.

First of all, there are various approaches that try to implement a computational forward strategy: simulate the chain and observe whether it has reached stationarity. This sort of observation remains shaky because there might be relevant areas that the chain has not yet visited, or not visited with significant frequency. Maybe waiting twice as long will change the observed distribution significantly. As we mentioned previously, the effectiveness of MCMC relies essentially on how quickly Markov chains converge to their stationary distribution, sometimes called fast mixing. The speed of mixing is relative to the complexity of the space the random walk has to explore. The important question is: Exactly how quickly does the chain actually converge?

Answering this question is crucial for any assessment of MCMC results. Fast mixing and the rate of convergence have been identified as an important research topic being tackled by some of the most prominent researchers in stochastics and statistics.[28] Despite a growing number of results and insights, there is still a large lacuna regarding the behavior of chains that move in large continuous spaces, as is typical with Bayesian posteriors. In fact, this is the downside of MCMC-enhanced modeling flexibility: the menagerie of MCMC-enhanced models is growing, whereas knowledge about convergence speed is still lacking. There may be a chance for a mathematical theory to eventually provide such a footing; yet up to now, no strong results exist. Diaconis (2009, 195) reasons that the market may be populated by many applied MCMC algorithms that perform well, and that their careful analysis might present useful hints that would direct mathematical research toward why these algorithms behave so well—or toward why they do not. Diaconis has no illusions about how limited the range of mathematical accounts of the validation problem is. He is alarmed by the tendency to build excessively complex models for which, thanks to MCMC, the Bayesian machinery still works, whereas considerations (about MCMC) that could help regarding validation are largely missing.

This exemplifies a problem whose significance goes beyond the case of statistics and Bayesianism: namely, a technology-based mode of mathematical modeling pushes the limits of modeling so that questions of validation can be addressed only by quasiempirical means—that is, by observing the performance of the models. This state of affairs is endemic in computational modeling. Many researchers resort to a kind of quasiempirical forward strategy—that is, they explore via simulations how the model will behave under varying initial conditions. Carlin and Louis (2000), for an instance from Bayesian statistics, argue:

> The most basic tool for investigating model uncertainty is the sensitivity analysis. That is, we simply make reasonable modifications to the assumption in question,

> recompute the posterior quantities of interest, and see whether they have changed in a way that has practical impact on interpretations or decisions. (194)

This validation strategy—explore and observe variation in model behavior—can be found in many areas of computational modeling. It is characteristic of a field in which predictions are created in the iterative–exploratory mode. There is nothing wrong with these strategies; they just express that modeling happens under a condition of partial epistemic opacity in which model behavior is not controlled by clear-cut assumptions but rather by an assemblage of epistemic and instrumental components whose resulting behavior is adjusted.

7.4 Software

Software plays an important role in the upswing of Bayesian methods. A revolution from the perspective of professional mathematicians and statisticians might not necessarily have great impact on the methods practitioners use. As with any other instrument, the (perceived) quality of the instrument has to attract and hold the interest of an array of potential users and, furthermore, must be usable given their level of expertise. Software packages have been—and still are—key for distributing the iterative–exploratory approach inherent in MCMC that forms the computational backbone of Bayesian modeling.

Such software is not a neutral framework because the options it offers and the algorithms it implements tend to steer in whatever directions statistical practices move (cf. Mira, 2005). Bayesian software has had two major effects: one the flip side of the other. Due to its usability, together with the easy accessibility of networked computers, it has triggered a stunning distribution of Bayesian modeling far beyond the ranks of those who had a Bayesian inclination before the 1990s turn. At the same time, many of these novices in statistical modeling are attracted by the software's capacity to deal with more complex models rather than by the standard rationale of Bayesianism.

When the great potential of MCMC began to become manifest, MCMC pioneers such as A. F. M. Smith (1988) realized that a software package was the missing ingredient that could turn Bayesian modeling into a widely used approach. This was exactly what David Spiegelhalter and his coworkers at the MRC Biostatistics Unit in Cambridge (United Kingdom) were developing. In 1991, they rolled out the BUGS program (short for Bayesian Statistics Using Gibbs Sampling). Freely available, it popularized Bayesian modeling tremendously. BUGS acted as a platform for Bayesian modeling by generating code for MCMC-based analyses of models that users could specify (see Gilks et al. 1994; Thomas et al. 1992). It featured uncertainty propagation in graphical structures; but the main point, of course, is that modelers could use the software to compute a posteriori distributions of their models without having to master the mathematics of MCMC.

Not much later in 1996, now under Nicky Best who had changed from Cambridge to Imperial College, London, the descendant WinBUGS was published—a version running under Windows, reflecting the growing demand from the side of users who had no connection to special computing facilities but worked on (relatively small) desktop computers.[29] WinBUGS acted as an efficient popularizer, enlarging the variety (and complexity) of possible models as much as the variety of users. Textbooks such as Ntzoufras (2009) guided readers into using WinBUGS, with the selling point being that this free software "could fit complicated models in a relatively easy manner, using standard MCMC methods" (xvii). The entire book is devoted to WinBUGS, but time runs quickly in software development. By the end of the 2000s, the BUGS program had been turned into the open-source code OpenBUGS that is very similar to WinBUGS but also runs on Linux, Apple, and other Unix-related operating systems.[30] Importantly for computational modelers, OpenBUGS can be run from R and from SAS—that is, from the most common platforms of statistical analysis—thus creating a software environment for statistical modelers.

There is a plethora of packages that come into play on different levels. Some scientists use MCMC methodology by interfacing their data with a (more or less) complete tool for analysis like the BUGS family offers.[31] The BUGS family is not the only type of software package. Others invest work in developing their own customized MCMC simulations using software packages such as Mathematica or MathLab more as a generic tool kit. One important feature of "complete tool" software such as BUGS is the way one can make use of it. It not only provides a graphical interface but also comes with a book of examples. Hence, users do not have to learn how it works—that is, how to specify their model case. Instead, they can build directly on particular examples. As Carlin puts it: "You don't read the manual; instead, you find the example that most nearly matches your situation, copy it, and modify it" (Kass et al. 1998, 94; cf. also Carlin [2004]).

Now, however, the flip side comes into play. MCMC has unresolved issues with convergence as we have seen earlier. Standard software has no guardrails that would prevent users from ignoring this issue. Jeff Gill (2008), for instance, recapitulates the enormous success of software packages in solving the needs of modelers but also warns:

> Unfortunately, these solutions can be complex and the theoretical issues are often demanding. Coupling this with easy-to-use software, such as WinBUGS and MCMCpack, means that there are users who are unaware of the dangers inherent in MCMC work. (xx)

The OpenBUGS website issues a "Health Warning": the programs are reasonably easy to use and come with a wide range of examples. There is, however, a need for caution. Knowledge of Bayesian statistics is assumed, including recognition of the potential importance of prior distributions; and MCMC is inherently less robust than analytic statistical methods. The fact that there is a (largely) unknown level of uncertainty should sound an alarm. However, there is no built-in protection against ignoring this fact.

Not surprisingly, the convergence problem is a matter addressed in some of the available packages. Interestingly, because a mathematical solution of the validation problem is out of reach, the software resorts to heuristic strategies to explore model behavior. The software AWTY (Nylander et al., 2008) provides a case in point. The acronym expands into "are we there yet?"—that is, has the chain reached equilibrium? The program is made for graphical exploration of convergence in the special case of Bayesian phylogenetics.

One can ask to what extent the lack of built-in guardrails poses an actual problem in statistical practice. This is hard to judge. It is not at all unlikely that practical methods are valid, although they cannot be fully justified mathematically. In lieu of a reasoned judgment, I can only offer an impression: it looks as if the standards of what counts as sound methodology are beginning to change. They are moving away from a mathematical paradigm tied to proof to a computational paradigm in which skillful modification is the key.

The wide uptake of Bayesian methods reflects the social organization of the field. The number of users was able to grow dramatically because—thanks to the software—these users do not need to be experts in either statistics or the mathematics of MCMC. Bayesian methods have become a pragmatic and flexible tool in statistical practice. At the same time, this flexibility leads to an erosion of the original rationale: whether frequentist subparts are utilized or whether priors express a meaningful subjective stance is not per se important if the machinery works.

7.5 Prediction and Pragmatism

Bayesian approaches are a success story in statistics that began in the 1990s. We have argued that this story pivots on the co-development of computational methods and a class of models that, when working together, made predictions possible. From a methodological perspective, MCMC was the key factor; from a social perspective, the widely available software that runs on networked computers was a key contribution to success in practice. In a nutshell, Bayesian statistics evolved into an exemplar of the exploratory–iterative culture of prediction. Importantly, this evolution affected Bayesian rationality in an important way. Namely, the interpretation of priors changed from an expression of beliefs held prior to new evidence into an adjustable parameter that can be manipulated flexibly by computational machinery—an auxiliary for exploratory iteration. This change transforms Bayes's rule from a principle motivated and justified by rationality into a tool efficient in making predictions. Thus, in the case of Bayesian statistics, we see a coevolution of computing technology, the exploratory–iterative mode of prediction, and the conception of rationality.

This has the potential to fundamentally affect the philosophy of statistics. The principled interpretation of Bayes's rule has been the major bone of contention. Statisticians are aware of this transformation, not least because elements that formerly counted as incompatible now come together in predictive practices. How the new situation should be captured conceptually is not yet clear. Some scholars want to restrict the title "Bayesian" to approaches that stick to the Bayesian principle. They are critical of the newer prediction-oriented approaches. Others, who still perceive themselves as Bayesians, side with prediction-making.[32] Next is a sample of responses from statisticians.

According to Bradley Efron,[33] classical frequentist and Bayesian approaches work together and mutually *complement* each other in computer modeling. Especially when analyzing large amounts of ("big") data—according to Efron (2005)—it is often hopeless to construe priors in a subjective way. Sander Greenland (2010) argues that Efron's stance on the mutually complementing virtues is not correct and that it would be better to use the term "ecumenism" to describe how statistical methods come together.[34] He traces this back to G. E. P. Box's (1983) plea for ecumenism. Despite its prominent advocates—according to Greenland—ecumenism has not yet had a large impact on the teaching or practice of statistics.[35] Robert Kass is another prominent statistician who reflects on the ongoing changes in a conceptual way. He advocates what he calls "statistical pragmatism," a position that sees modeling as the core activity (Kass 2011). He makes a careful attempt to sketch the common ground between Bayesian and frequentist positions regarding how statistical models are connected with data. Thus, the dynamics of computational modeling seem to be a uniting feature of formerly separated camps of philosophy of statistics: "The loyalists of the 1960s and 1970s failed to realize that Bayes would ultimately be accepted, not because of its superior logic, but because probability models are so marvelously adept at mimicking the variation in real-world data" (Kass, cited according to McGrayne [2011, 234]).[36] Steven Goodman (2011) disagrees because Kass's pragmatism looks like a mere truce rather than a new foundation. Also commenting on Kass, Hal Stern (2011) worries "more broadly that pragmatism might appear to reinforce the notion of statistics as a set of techniques that we 'pull off the shelf' when confronted with a data set of a particular type" (17). Finally, Andrew Gelman (2011, 10) observes that this pragmatism, though thriving on the flexibility of methods to obtain calibration between model and data, is still objective. In sum, notions such as complement, truce, ecumenism, or pragmatism show that statisticians grapple with reflecting on what happens in practice and whether this makes a discussion about the foundations dispensable or, on the contrary, downright demands it. Although there is no settlement in sight, one message is apparent: the exploratory–iterative culture of prediction is bringing forth a new discussion about the foundations of statistics.

8 Engineering Thermodynamics

Thermodynamics features prominently in both science and engineering. First, it is a paradigmatic example of a fundamental and broadly applicable scientific theory. Its main laws—the first is about conservation of energy and the second about the increase of entropy—instantiate gems of a rational viewpoint (in the sense of the rational culture of prediction introduced in chapter 2). Second, thermodynamic processes are at the core of many engineering feats. The steam engine[1] is linked particularly to thermodynamics because it triggered its evolution. It is precisely for this reason that it is so prominent in the history of science and engineering—the steam engine predated the theory, hence defying claims that engineering knowledge would rest on, or derive from, scientific knowledge.

This chapter examines engineering thermodynamics, a field that has received surprisingly little attention from history and philosophy of science. Thermodynamic engineering is concerned with designing (chemical) processes, and it is fundamentally about prediction. Engineers ask questions such as: What will be the pressure in a tank of a given volume if one loads it with a given amount of a given mixture at a certain temperature? To answer such questions, engineers use equations of state (EoS) that express the relationship between pressure, temperature, and further variables. More generally, they are used wherever engineers need to characterize properties of materials, and they are standard mathematical tools for prediction.

The most basic EoS is the ideal gas law, formulated by the engineer and physicist Benoît Clapeyron (1799–1864) in 1834. It holds for all substances, but only if the molecules do not interact—that is, for the "ideal gas." This law is part of the *rational* culture of prediction (see chapter 2)—it claims universal validity but ties prognostic value to idealizing assumptions. At the same time, this law is also part of the *empirical* culture of prediction because Clapeyron synthesized mainly empirical results.[2]

In other words, today's thermodynamics engineers construct EoS in a way that merges both cultures. They combine mathematical formulation and predictive capacity with adaptation to empirical results. But a given EoS can be used only *if* this equation captures the complexity of the target in an adequate way *and* if it is tractable. Hence, developing EoS is a difficult and limited theoretical endeavor. The iterative capabilities of computational modeling changed the game, we argue. More precisely, engineers and scientists develop and use predictive EoS through utilizing adjustable parameters in exploratory modeling.[3] This sort of exploration became feasible with the computational infrastructure of easily accessible computers—it characterizes the exploratory–iterative culture of prediction. In fact, we argue, making predictions is based on a merger of all four cultures discussed in this book—the rational and the empirical as well as the computer-based iterative–numerical cultures of prediction. Moreover, we claim that the recent exploratory–iterative culture is a crucial ingredient because it features adjustable parameters.[4]

In section 8.1, we shall describe EoS and how they produce predictions. The wide variety of EoS form a structure resembling a tree with the ideal gas law as its root. This law is the simplest EoS. From there, branches with other EoS grow toward complexity, mainly through parameterization. In many relevant circumstances, the predictive quality of the EoS depends crucially on the number and type of its parameters and on the way in which they have been adjusted. However, "climbing up the tree"—that is, working with more complex EoS—is generally not feasible without simulation modeling. Hence, the second section starts with a systematic schema of simulation modeling and highlights the decisive role of a feedback loop through which modeling and experimentation are connected. This loop is the key methodological ingredient for exploratory modeling—adjusting parameters thrives on it.

With the framework of thermodynamics (section 8.1) and simulation modeling (section 8.2) in place, section 8.3 presents the centerpiece of our analysis: a closer look at different types and functions of parameters. Parameterization is a key element because it effectively addresses three types of gaps. Typically, adjustable parameters are used to work around gaps in mathematical tractability and gaps in which theoretical knowledge is missing *at the same time*. And because parameterization is done over the course of simulation modeling, it also bridges the gaps in the software. Thus, parameter adjustment is a crucial tool for producing a prediction that draws on—and convolutes—theoretical, empirical, and computational resources.

In section 8.4, we shall wrap up our findings and argue that utilizing adjustable parameters entails a mixed bag of implications. In particular, EoS wield predictive power at the cost of their explanatory capacity.[5] Merging theoretical, empirical, and technological resources is often without alternative when aiming at predictions. This merging has become possible because technology and mathematization are intertwined in a way that we call the exploratory–iterative mode of prediction. In short, thermodynamic engineering leverages an exploratory–iterative culture that draws on resources of different cultures and merges them in a way that aims at prediction.

8.1 The Branching Tree of Equations of State

Our thermodynamic examples will focus on EoS for fluids, while leaving solids aside.[6] The best known EoS is that of the ideal gas:

$$p\,v = R\,T \tag{1}$$

where p is the pressure, $v = V/n$ is the molar volume (volume per mole of substance), and T is the temperature measured in kelvins. All these quantities are measurable in classical experiments. R is a universal constant (8.314 J mol^{-1} K^{-1}). It should be noted that R was established on the phenomenological level—that is, it started its career as an adjustable parameter but later led to theoretical insights into molecular thermodynamics.

All substances fulfill equation (1) if the density $\rho = 1/v$ is low enough (or the molar volume v is high enough). Equation (1) is a showcase of a universal law (cf., e.g., Woody [2013]), carrying the idealized nature in its name; and this law is a special case for a much broader concept of EoS. Such an equation is a function describing how the quantities p, v, and T relate to each other for a given amount of a given substance. The general form of such an EoS is accordingly:

$$f(p, v, T) = 0. \tag{2}$$

In equation (2), a pair of independent variables can be chosen (e.g., p and T), and the third (dependent) variable (then v) can be calculated. In the low-density limit, the function f is given by the simple equation (1): $p\,v - R\,T = 0$.

At higher densities, however, the interactions between the molecules start playing a role, and hence, the functional relationship becomes much more complex and also specific to the substance. Hence, a tree of EoS grows from equation (1) and branches toward more complex EoS that aim to extend the region of applicability of EoS beyond the low-density limit of the universal equation (1). When climbing up the tree of EoS, the first two branches are the van der Waals EoS:

$$p = \frac{RT}{v-b} - \frac{a}{v^2} \tag{3}$$

and the virial EoS

$$\frac{pv}{RT} = 1 + B\frac{1}{v} + C\frac{1}{v^2}. \tag{4}$$

The researchers who introduced these equations, J. D. van der Waals (1837–1923) and H. Kamerlingh Onnes (1853–1926), were awarded Nobel prizes in physics in 1910 and 1913. These equations, though both with strong theoretical roots in physics and mathematics, contain adjustable parameters—namely, a and b in equation (3) and B and C in equation (4). The nature of these parameters changed over the development of thermodynamics. For example, the theory behind equation (4) yields that B and C are functions of the temperature but not of the pressure. In the original version of equation (3), a and b were numbers. However, in later versions of equation (3), a is considered to be a function of temperature. Adjusting functions is obviously more flexible than adjusting numbers. These parameters are needed to account for the individuality of different fluids (e.g., oxygen or nitrogen). In other words, numbers that are found for the parameters a and b in equation (3) differ depending on whether one is studying oxygen or nitrogen.

A major asset of EoS lies in their ability to describe properties of fluids with great precision. Often, parameterizations of p can be tested against v and T data available from laboratory experiments. A good EoS will match these experimental data over a wide range of conditions. However, it follows from thermodynamic theory that EoS can do much more: they contain information about many other interesting quantities. If an EoS describes both gaseous and liquid states, it also describes boiling and condensation. For instance, equation (2) also determines the so-called vapor pressure curve describing how the boiling temperature depends on the pressure. These results can be compared to experimental data as well. Another instance would be the heat of vaporization that can also be obtained from equation (2).

An EoS can be modified in two ways: first, by structural modification, which would lead to a new EoS—that is, a new function *f*. In this way, equations (3) and (4) emerged from (2). Second, the parameterization of one particular *f* can be changed—that is, by assigning values to parameters of the function such as choosing *a* and *b* in equation (3) to describe a given substance.

The number of EoS in the literature is enormous. Just counting only the different mathematical forms of equation (2) adds up to several thousand equations. Considering the specifications for particular substances or mixtures of substances would further increase that number dramatically. Not only has the number of EoS has soared but also the number of adjustable parameters used in EoS. Whereas there are also new EoS with only a few adjustable parameters, others have more than fifty (e.g., Span et al. 2000). Groundbreaking new ideas, which would lead to the creation of new class of EoS, are comparatively rare. The proliferation of EoS is created more by introducing new parameterizations for a function *f* that is already in use, and it is driven by the demand to be able to use EoS for very specific predictions such as for a certain class of fluids in a certain range of conditions. Clearly, the use of computers has enabled such profusion. In precomputer times, handling only very few parameters was already demanding. The computer and its numerical and iterative capabilities changed the situation. Parameterizing EoS—adjusting them to the circumstances of interest—has turned from a tediously time-consuming endeavor done by specialists into a commodity that is in easy reach for many users.[7] In this way, adjusting parameters is an efficient instrument with which to gain predictions from EoS.

In this book, we have distinguished between two computer-related cultures of prediction: the iterative–numerical mainframe culture from the exploratory–iterative culture. The latter is more recent (starting around 1990) and tied to easily available networked computers. This chapter largely ignores the former one because adjusting parameters has a strong exploratory element. Bibliometric data from the Web of Science corroborate the time frame. The number of papers that work with EoS started to grow significantly around the 1990s (see figure 8.1). Additionally, the data document the growing ties between EoS and engineering. Whereas in 1985, physics publications dominated, and the share of publications classified as "engineering" lingered

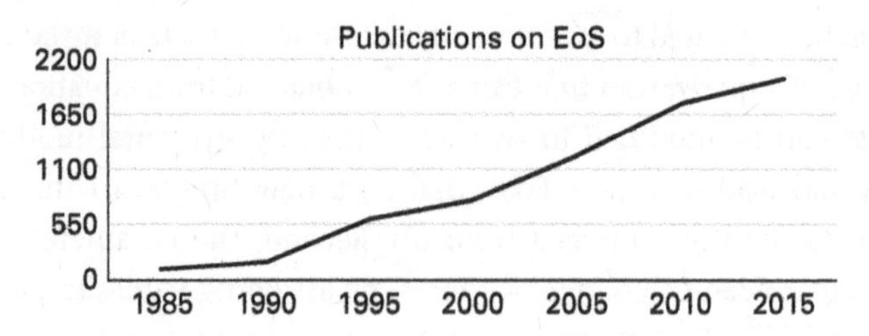

Figure 8.1
The number of papers per year in the Web of Science databank with "equation of state" in their title or abstract, in five-year steps and linear interpolation. Courtesy: authors.

around 25 percent, this share had risen to 75 percent by 2015.[8] This suggests that the usefulness of EoS as tools for prediction in engineering set in around 1990.

8.2 Simulation Modeling, Experiments, and Feedback

In the present account, the term model[9] is used in the sense of a mathematical model that aims to depict certain aspects of physical reality such as an EoS that describes the vapor pressure curve for a given substance. This model is considered to be embedded in some kind of theory that provides a frame not only for the modeling but also for carrying out experiments on physical reality that deliver outcomes that can be compared with those of the model. Simulation models are models that are evaluated by carrying out (computer) simulations as opposed to models that can be evaluated by other means such as pencil and paper. Figure 8.2 gives an overview of the picture on which the following discussion is based.[10]

Modeling is an inevitably indirect procedure. Modelers might target a quantity x^{real} in the real world such as the vapor pressure curve.[11] The corresponding entity in the model is x^{mod}. Because the model is too complex to be evaluated directly, it is implemented on a computer and simulations are carried out. These simulations yield a quantity x^{sim} that can be compared eventually to the results of experimental studies x^{exp}. In general, neither x^{real} nor x^{mod} can be known; only the corresponding properties x^{exp} and x^{sim} are known and can be compared.

Both are known through types of experiment: one from "below" that provides measured values and the other from "above" that provides simulated

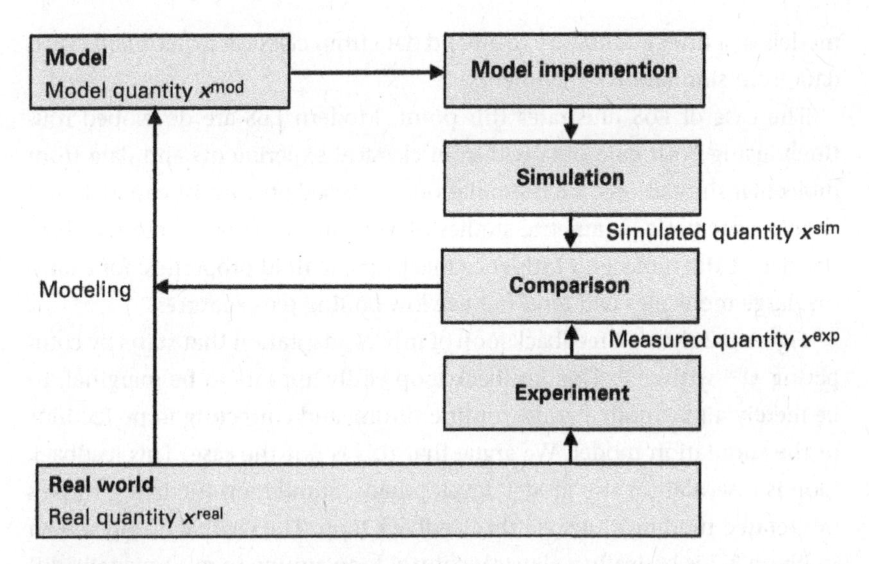

Figure 8.2
A schema of simulation modeling including a feedback loop. Courtesy: authors.

values.[12] The latter are often called computer experiments or numerical experiments. However, we prefer using the term "simulation experiment" (or briefly just simulation) here. Such experiments are used to investigate the behavior of models. Importantly, it is only through simulation experiments that relevant properties of simulation models can be known.[13] This has two immediate and important consequences: first, simulation experiments are unavoidable in simulation modeling. Second, when researchers construct a model and want to find out how modifications of the current version perform, they will have to conduct repeated simulation experiments.

The second variety ("from below" in figure 8.2) is the experiment in the classical sense. When comparing simulations to their target system, such classical experiments will usually provide the data for the comparison. However, the situation becomes complicated in an interesting way because of the growing influence of simulation on these experiments. Debate is just beginning on how computer use is changing the face of measurement and experimentation (cf. Morrison [2009; 2015]; Tal [2013]). Nonetheless, this debate is already showing that the clear-cut separation between the two types of experiments as suggested in figure 8.2 is oversimplistic. This is underpinned by the fact that when developing new models, researchers use models of a finer granularity to amend data from classical experiments with data from simulation experiments.

The case of EoS illustrates this point. Modern EoS are developed routinely using both data obtained from classical experiments and data from molecular simulations. Such simulations are based on atomistic models and enable systematic parametric studies of how atomistic parameters such as the size of the molecules influence macroscopic fluid properties: for example, large molecules will tend to have low boiling temperatures.

Figure 8.2 shows a feedback loop of model adaptation that starts by comparing x^{sim} with x^{exp}. This feedback loop easily appears to be marginal, to be merely a pragmatic handle for fine-tuning and correcting imperfections of the simulation model. We argue that this is not the case. This feedback loop is essential for the model development. Simulation modeling thrives on iterated modifications via this feedback loop. The feedback loop shown in figure 8.2 is basically a classical control loop aiming to minimize the differences between a variable (here: x^{sim}) and a set value (here: x^{exp}).[14]

Within the feedback loop, repeated comparisons of the results of the two types of experiment guide the modeling process. The new versions of the model obtained in the loop are explored via simulation experiments and compared to results from classical experiments. Hence, this reveals a cooperation between both types of experiment that is the core of model development via the adjustment of parameters.[15]

Parameterization schemes are architectures that guide actions of this type.[16] They can be considered as a sort of auxiliary construction that is used intentionally to deal with missing knowledge and the inaccuracies of existing knowledge. The simulation model is designed to contain parameters that can be adjusted over the course of further development. Hence, parameterization schemes supply flexibility to a model. Even a model whose structure represents the structure of the target only poorly can give fair representations of x^{exp} after a suitable adjustment of parameters. However, this implies that conditions regarding technology (computational power), social organization (access), and economy (costs) are fulfilled. Hence models, methodologies, and technologies coevolve.

8.3 Adjusting Model Parameters: A Closer Look

In the remainder of this chapter, we shall focus on the role of adjustable parameters—such parameters that cannot be eliminated from simulation modeling. We discuss the reasons for using adjustable parameters and also their (un)intended effects in more detail.

The Control Loop Machinery

Figure 8.3 presents a simple scheme of a feedback control loop—a common element in systems theory. The process model aims to describe a certain set of quantities ***y*** that we shall call output variables. The output depends on the input, which is described by another set of quantities ***u***: the input variables. Both ***y*** and ***u*** belong to the quantities that occur in the model. We shall call the latter model variables ***x***. The set of the model variables ***x*** may contain quantities that are neither input nor output (i.e., internal variables). The question of which subset of ***x*** is considered as input and which as output may depend on the application. In the models considered here, ***y*** and ***u*** describe properties of the target system. Ideally, ***y*** is a measurable quantity and ***u*** can be set in experiments.

Alongside the input variables ***u***, many models require the specification of model parameters ***p***, as shown in the case of EoS. By cleverly setting the model parameters, the quality of the model can be improved in terms of its output ***y***, and therefore especially in terms of predictions.

Adjusting the parameters involves some kind of optimization procedure. The goal of optimization is to improve the fit between the model output ***y*** and some reference data, usually experimental data y^{exp} (see figure 8.3). Its outcome is a parameter set that becomes an inherent part of the model. This parameter set, and hence the model itself, will depend not only on the type of reference data chosen for the comparison but also on the way the comparison is carried out.

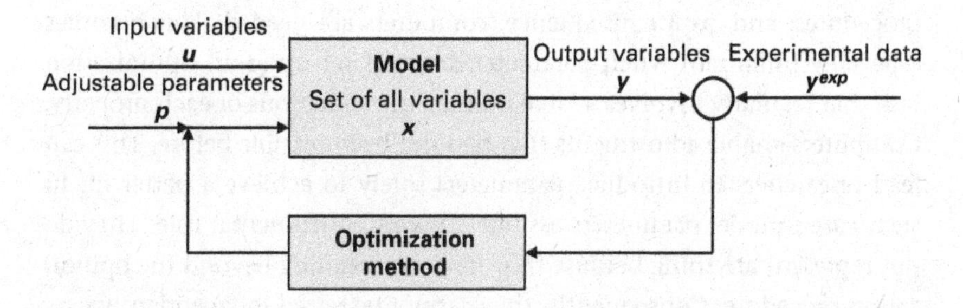

Figure 8.3
Parameter adjustment in model development as a control loop that uses some optimization method. Courtesy: authors.

We are not rigorously presupposing some elaborate formal algorithm for optimization. A simple trial-and-error method is also eligible for the "method." Nonetheless, we wish to point out that mathematical optimization methods extend far beyond what can be handled by simple trial and error. Such methods may present black boxes to modelers when, for instance, a numerical solver is applied. Nonetheless, such solvers can deal with a surprisingly large number of parameters. What had been inadmissibly tedious when having to be done by hand has now become feasible thanks to automated iteration by the computer.

Figure 8.3 highlights that adjusting model parameters aims to produce predictions. Numerical optimization targets overall performance for a given data set and performance metrics. Therefore, practical values guide the adjustment more than epistemic ones. Furthermore, it shows that the applicability of a model is determined not just by its form. It is also influenced strongly by the choice of the dataset on which its parameters were trained.

The van der Waals equation (3) can be used as an example to illustrate this point. The input variables may be chosen as the temperature T and the molar volume v; and one may be interested in the result for the pressure p at these chosen conditions. The calculated result will depend on the choices made for the parameters a and b. Obviously, if some p, v, and T data points are available for a given substance, the parameters a and b can be adjusted to these data. Additionally, the results obtained for a and b will depend not only on the choice of the dataset to which they are fitted but also on the way they are fitted.

To parameterize EoS, different types of data are used (e.g., alongside p, v, and T data, also data on vapor pressures, data on the critical point of the fluid, or caloric data). Calculating such properties requires numerical procedures; and, as a consequence, computers are needed. This becomes especially important when parameterization is set up as an optimization task that regularly involves a large number of evaluations of each property. Computers enable adjustments that had not been feasible before. This can lead researchers to introduce parameters solely to achieve a better fit. In such cases, model parameters assume a purely instrumental role. They do not represent anything because they have no meaning beyond the optimization procedure. Consequently, they cannot be tested independently.

On the other hand, in the case of EoS, there is an obvious need to adjust parameters. With only a few exceptions, science is not yet capable of predicting properties of real fluids from first principles. At present, such predictions are limited more or less to calculating ideal gas properties from Schrödinger's equation. In all other cases, models describing such properties must be trained with some experimental data. The way to do this is via adjusting model parameters. We would like to point out that the control loop does not just operate with an instrumental logic but combines theoretical and empirical reasoning. Equations of State such as equations (3) and (4) are far more than some arbitrary mathematical form that is fitted to data. They do not just contain equation (1) as limiting case. Moreover:

- The B parameter of equation (4) can be related directly to intermolecular pair interactions and was for a long time the most important source of quantitative data on them.
- The simple equation (2) predicts the existence of phenomena such as critical points or the metastability of fluid phases and relates these to other fluid properties in a consistent way.

These examples highlight the power of thermodynamic theory and illustrate the benefits of combining theory and experiment: the powerful theory can be exploited only if an EoS is available that has been adjusted to some experimental data. Both parts together—that is, a combination of theoretical structure and pragmatic flexibility—are crucial for attaining predictions.

An illustrative case in point is how EoS can be used to describe mixtures of substances. One has to find expressions for the parameters of the equation (such as a and b in equation [3]) that hold for the mixture. These mixture parameters are usually calculated from the corresponding pure component parameters and the composition of the mixture parameters via so-called mixing rules. With the exception of the mixing rules for the parameters of equation (3), which can be determined rigorously from the principles of molecular thermodynamics, mixing rules for EoS are essentially empirical. They contain parameters that usually have to be adjusted to mixture data (Wei and Sadus 2000). Nevertheless, they can be submitted to some tests that may be either of a logical nature (i.e., if a pure component is split up formally into two identical subcomponents, the pure component result should also be obtained from the mixture model) or based on theoretical findings in thermodynamics such as those from molecular thermodynamics mentioned previously. It is known that mixing rules that have failed in both logical tests and tests from molecular thermodynamics can nevertheless prove to work well in practice when the parameters are adjusted suitably. From a pragmatic standpoint, they may even perform better than theoretically sound mixing rules (for examples, see Mathias et al. [1991]). These are cases in which the overall parameterization, including all parameter assignments, gives a prediction that can be tested, whereas single parameters cannot be tested independently.

Proliferation of Model Variants

Model variants proliferate along the branches of the EoS tree. There is an enormous number of functional forms of EoS, choices of the parameters in these forms, and, on top of this, choices of the numbers for the open parameters for some given fluid. In fact, computers have opened the gates to this proliferation. The possibility of easily creating and checking variants of some model against empirical data is, at first sight, a positive development. On closer inspection, however, the picture changes: first, the plethora of variants of a given model will rarely have epistemic value. Second, they will create an obstacle for anybody wanting to use that model. Which variant to choose? By facilitating the creation of sprawling mutations of models, computers contribute to the fragmentation of research and even compromise its actual applicability.

The van der Waals equation (3) can be reconsidered as an example. Developed in 1873, there are now more than four hundred equations of state (so-called cubic EoS) that can be considered to be variants of that single equation (Valderrama 2003). Although this grants, of course, enormous credit to the groundbreaking work of van der Waals, it is also a source of concern in the community of researchers. The variants can hardly be classified on theoretical grounds. Instead, this is replaced by historical (When were they developed?), sociological (How well are they received?), or pragmatic (What practical benefits are offered?) arguments and classifications. Some very successful variants are widely used, and some older versions certainly have technical drawbacks, but there is also a plethora of variants that are very similar. Many of these have been used only by the group that originally proposed the equation. Thus, the exploratory–iterative approach to modeling that developed together with an infrastructure of highly available computers has social repercussions of an almost paradoxical character. The networked infrastructure encourages the fragmentation of modeling groups. This has been captured nicely by the computational chemist Daan

Frenkel (2013) in his paper on the "dark side" of simulations: "In the past, we had to think about the role of simulations because they were expensive, now we have to think because they are (mostly) cheap" (footnote 1).

Need to Adjust Model Parameters

What makes this parameterization problem so endemic and, in a sense, unavoidable? In general, any mathematical model presents an idealized version of a real-world target system. There is always a greater abundance in the target system than in some mathematical model equations. Hence, there may be both known and unknown properties of the target system that should be, but have not been, included in the model. Leaving open some model parameters and adjusting them to experimental data can be considered as a pragmatic remedy: the model meets the diversity of the target system through strategic flexibility.

Even if all the properties of the target system that exert an influence were to be known, it might still be prohibitive to account explicitly for their influence in the model. There may simply be a lack of theories, or existing theories might be so complex that they would make the model intractable. Adjustable parameters are of prime importance in this context. They make it possible to use simplified but tractable models. Such models may be related only loosely to the target object and may be obvious oversimplifications. But leaving open some parameters in such models and adjusting them in ingenious ways can make them work. This is part of the modeling activity in applied science and engineering. In the vast majority of cases in engineering and science, the choice is not between having some model without adjustable parameters and having one that contains such unconfirmed elements. Rather, the choice is between having a model with adjustable parameters and not having a (relevant) model at all.

Again, the van der Waals equation (3) can be used as an illustration. In equation (3), the parameters a and b have a physical meaning. They are associated with attractive and repulsive interactions between particles. It is well known that there are many different types of attractive forces that are all lumped together in the parameter a. It can, hence, be considered as an "effective" parameter—that is, a parameter that describes the influence of an entire class of physical phenomena (attractive forces) within a given model. In addition, the parameter b can be considered as such an effective parameter describing repulsion. Despite the crude simplifications in the assumptions on the intermolecular interactions, the van der Waals equation and its mutants have been extremely successful in describing real fluids. There are two main reasons for this: first, the structure of the equation (which comes from theory) is able to qualitatively reproduce many important features of the behavior of fluids such as the coexistence of vapor and liquid at certain conditions or the ideal gas-limiting behavior. Second, the equation contains the adjustable parameters that create flexibility and thus can alleviate shortcomings of the theory. Both factors—adequacy and flexibility—act together.

Parameters with and without Independent Physical Meaning

In principle, any variable in a model can be used as an adjustable parameter. However, two cases should be distinguished: whether or not the parameter has an independent physical meaning. Independent physical meaning is used here in the sense that there is a physical interpretation outside the context of the parameter fitting.

The van der Waals equation again illustrates this case: assume its parameters are fitted to the experimental p, v, and T data of some liquid. On closer inspection of equation (3), one finds that the liquid density at high pressures is determined by the b parameter (since $v = b$ for $p \rightarrow \infty$). Hence, one can interpret the b parameter physically as describing the liquid density at high pressures. We consider this to be an interpretation in the context of the parameter fitting here, and hence not as an independent physical interpretation. However, as stated previously, by virtue of the derivation of the van der Waals equation, the b parameter has a deeper meaning. It describes repulsive intermolecular interactions. These obviously become very important in liquids at high pressures in which the distances between the particles in the fluid become very small. Repulsive interactions can, in principle, also be determined independently, namely, from quantum chemistry. Unfortunately, the derivation of the van der Waals equation is based on such crude simplifications that there is no way to relate or predict the b parameter from independent sources of information such as quantum chemistry.

The above statement shows different things: whereas it is fair to say that b is related to repulsive interactions, there is no way to establish such a correlation quantitatively on a theoretical basis. An important consequence of this is that the numbers for b obtained from the fitting procedure should not be overinterpreted as carrying useful quantitative information on the repulsive interactions. This cautious conclusion is supported by noting that the numbers obtained for the *b* parameter of a given real fluid will depend strongly on the choice of the dataset used for the fitting. Nevertheless, it is obviously a merit of the van der Waals equation that it delivers structural insight into the importance of certain interactions under certain conditions—in the present example, the repulsive interactions in liquids at high pressures.

In the general case, parameters either do or do not have an independent physical meaning. First, consider the case in which the variable used as a parameter has an independent physical meaning. By using it as an adjustable parameter, that physical meaning is initially abandoned. A number is assigned to that variable in the feedback loop based on pragmatic considerations about the overall model performance while disregarding the physical interpretation that the resulting number may have. However, one may try to recover the physical interpretation after the parameterization by comparing the result with some independent information on the property—if such information is available. Even if the result of this comparison is not promising, this does not compromise the usefulness of the overall model for predictions. However, such an outcome will shed a bad light on its explanatory power. On the other hand, it might turn out that the fit produces a parameter value that is "physically reasonable"—in other words, that meets some expectations based on considerations that were not included in the fit. This would be an indicator for the epistemic value of the model, even in the strong sense of predicting physical phenomena not only qualitatively but also quantitatively.

In the second case, the variable used as a parameter has no independent physical meaning. At first glance, that may seem to be trivial. One simply obtains some numbers from the fit, and there is no need or possibility to interpret the results for these numbers. All that can be done is to check the overall model performance. In practice, the typical situation is more complicated: parameters of models chosen for entirely mathematical reasons (e.g., coefficients of Taylor series expansions) may turn out to have a strong independent physical meaning. For example, equation (4) can be considered as a Taylor series expansion around the state of the ideal gas, and *B* and *C* are just the first two coefficients of that expansion. The theory of molecular thermodynamics shows that these coefficients are related directly to the energy of pair interactions in the gas. This means that parameters can lose and sometimes also gain a physical interpretation. Adjustable parameters help to *articulate* a theory toward applications.

Parameters in the Implementation

So far, we have discussed only model parameters. We have neglected the fact that the (theoretical) models often cannot be studied directly. They first have to be implemented on computers. Philosophers of science have pointed out that the path from a theoretical to an implemented simulation model involves a whole chain of models (Winsberg 1999), and that the discrete model might be in partial conflict with the theoretical one (Lenhard 2007). As a consequence, the implementation, which is a part of the feedback loop of modeling, will also influence the model parameterization. Aside from implementation errors, the differences between different implementations of one model are luckily often small enough to be ignored. Model parameters determined in one study are used regularly and successfully in other studies, even though the model implementations differ. However, there is no guarantee that this will be the case. Recent examination of molecular dynamics models has shown that different implementations of the same model will usually not yield exactly the same results (Schappals et al. 2017).

The parameters emerging in the model implementation are a greater concern. Prominent examples of such parameters are those used in the discretization of models or those used to control the numerical solvers that are part of the optimization procedures. Ideally, these parameters are chosen from ranges in which their influence on the simulation result is negligible (e.g., the grids used for the discretization must be "fine enough," sometimes even "not too fine"). But it may be very difficult to guarantee this.[17]

When such parameters exert an influence on the simulation results, they can be adjusted actively in the modeling feedback loop to improve the simulation results. This is much more problematic than adjusting model parameters because it is implementation dependent. After modelers have adjusted this kind of parameter, their results depend on the context of the specific implementation. Consequently, different modeling groups cannot discuss the model independently from its implementation. This situation may easily be misused by claiming misleadingly that the model is a success. However, such success cannot be attributed to the model alone but only to the model in conjunction with a deliberately tuned implementation.

Issues regarding the implementation are hard to communicate—or, at least, they have no place in current publication practices. On the other side, if parameters of the implementation are used as adjustable parameters, this threatens the testing of a model in a fundamental way. To the degree that model behavior depends on the concrete implementation, no other group of modelers can confirm or disconfirm simulation results—short of duplicating the implementation. Alas, in any concrete case, it is hard to determine on what sort of adjustments the predictions rest. Did the modelers fiddle about with the implementation? This information is rarely available. Addressing this question seems to be a matter of the modelers' ethos. Many practitioners observe that such an ethos might be desirable, although they agree that it is far from established in extant research practices.[18]

8.4 Prediction and Integration

Exploratory modeling thrives on the versatility of models. Without such versatility, the achievements of thermodynamic theory could not be brought to bear in concrete situations and for concrete properties. Our analysis showed that adjustable parameters are the tool to operationalize the flexibility of the models. This elevates adjustable parameters to crucial components of the models.

In traditional philosophy of science,[19] such parameters count as insignificant supplements or even also as an irritation to the theoretical core of a model. To the extent adjustable parameters (co-)determine model behavior, they influence, and possibly blur, the part of theory. Thus, they fall under the category of what philosophers of science have called ad hoc modifications. The take of philosopher Karl Popper was that ad hoc modifications are generally bad because they shield hypotheses against falsifying evidence.[20] Analysis of recent and historical examples has led philosophers of science to a nuanced viewpoint that has partly revaluated ad hoc measures. Various studies have shown convincingly that ad hoc modifications form an important part of scientific practice that does not undermine confirmation (see Bamford 1993; Friedrich et al. 2014; Worrall 2010).

Yet, the fact that adjustable parameters contribute to predictive success does not imply that such parameters will always be legitimate. Some adaptability is admissible; but how much of it is admissible without giving in to complete arbitrariness? According to Worrall, when accommodation is "reasonably foreseen" in the theory, it is "good ad hoc" and does not jeopardize confirmation, whereas modifications that merely react to observed shortcomings are "bad ad hoc." Alan Chalmers (2013) follows Worrall but adds an empirical twist: for him, ad hoc modifications are admissible as long as they are "independently testable" (126).

In our case, however, a main finding is that the parameterization cannot be separated from the model itself because it bridges multiple gaps simultaneously. Consequently, reaction to observed shortcomings is exactly the point of parameterization schemes, and there is often no independent testing possible that would focus on single parameters in a larger parameterization. In other words, adjustable parameters might be legitimate and might bring predictive success even if they are outside of what philosophy of science sees as admissible, "good" ad hoc modifications.

Two questions suggest themselves. The first is an epistemological one: Does this imply that confirmation of simulation results is doomed? Not quite. Validation of a model is limited due to the holistic nature of parameterization (Lenhard 2019); but within these limits, validation can be strict. Even if tests cannot address parameters in isolation, it is still possible to test the entire parameterized model along multiple dimensions. Testing a model along dimensions to which it has not been adjusted (such as at different temperatures or pressures or on different physical properties) is still a strong criterion, even if parameters are not tested independently.

The second question asks: What are the reasons why the philosophical discourse (exceptions allowed) is so strikingly incongruent to scientific and engineering practice? Of course, one reason is simply that this discourse started before the exploratory–iterative mode of prediction rose to prominence. Another reason, so we speculate, is the low-rank status of prediction. In many philosophical approaches, prediction counts as a sort of technical achievement, whereas explanation is seen as the more virtuous epistemological goal. And we have seen that working with adjustable parameters is oriented toward prediction, partly to the detriment of explanation (section 4.4). However, rising interest in engineering and applied science—and in their tools for prediction—will bring the issue of parameterization to prominence.

Our final point is that we observe a strong integrative contribution of the exploratory–iterative culture. Our analysis of adjustable parameters highlights these as central elements in simulation models. It would be misleading to think this happens at the cost of theory. On the contrary, the theory of thermodynamics is a resource that modelers can tap into for a broad range of practical applications exactly because adjustable parameters bring versatility. Thus, using adjustable parameters is a *condition* for attaining successful predictions. This finding resembles the cases of computational chemistry (chapter 4) and Bayesian statistics (chapter 7) examined earlier in this book—and displays engineering thermodynamics as a member of the exploratory–iterative culture of prediction. By integrating elements of older cultures of prediction, the new exploratory–iterative culture of prediction leads to new neighborhood relations between science and engineering. Mathematical modeling itself turns to an exploratory–iterative mode. In thermodynamics, as well as in computational chemistry and statistics, science and engineering approaches share important methodological features.

V Both Hybrid and Pure?

9 Conclusion

Our investigation has been a journey into the philosophy and history of predictions—however, not just any predictions but those made by engineers and scientists on the basis of mathematical tools. It is widely considered that this is what makes predictions scientific: they do not rely on guesswork or epiphanic experience but are based on knowledge—that is, on theories or models that take a mathematical form. Because predictions can be derived mathematically, it is tempting to think of them as direct functions, almost emanations, of scientific and engineering knowledge. Following this line of thought, the spread and success of predictions would simply mirror the spread and success of science. However, this book has taken a very different point of view: we did not accept prediction by mathematical means as a functional unit. Our goal was rather to bring to the fore how diverse practices of prediction have been (and are) as well as to show how rich (and challenging) the history of prediction proves to be.

All chapters contribute to two main findings about diversity and structure: the first is that predictive practices are diverse and struggle to assert themselves against competing predictive practices. How predictive goals are formulated, the conditions of what counts as a prediction, and the varying ways of organizing the predictive endeavor—they all change. Notably, these changes happen at the same time and mutually influence each other—that is, they combine to form a picture of coevolution. The evolutionary perspective on prediction revealed how rich the topic is. It is not the conclusiveness but the *diversity* of predictive practices (with mathematical means) that is enlightening.

The second main finding balances the first by adding structure. We classify the nexus between epistemology, mathematization, technology, and social organization into *four cultures of prediction*. These cultures do not displace each other but rather coexist—a finding based on the longue durée perspective proposed in this book. The rational and the empirical cultures established themselves from early on in the history of modern science. The tension between the rational and the empirical modes accompanied prediction from the seventeenth century onward, though in constantly renewed forms. The iterative–numerical and the exploratory–iterative cultures are both linked to computer technology. The former has forerunners in precomputer calculating instruments, and it matured with mainframe machines, whereas the latter is tied to easy access and availability—that is, conditions present from around 1990 onward.

The significance of these cultures is evident from the breadth of answers given to the following two questions: What counts as a prediction? And when does it count as a predictive success? As the chapters have shown, answers to these questions are highly culture dependent—and often counterintuitive to members of other cultures (including the authors and probably also most readers of this book). For more than a century, predicting the trajectory of a projectile was considered a triumph of mathematized science, regardless of whether the correctness of the prediction was, or indeed could be, in any way confirmed empirically. In the late nineteenth century, engineers such as Bach, who wanted to bring empirical components to mathematization, argued that the predictions of an older, rational—and deeply mathematized—culture in engineering were not successful. However, this clashed with established opinion. The hybrid culture advocated by Thurston, Bach, and other engineers required new standards of success to become established with the help of engineering laboratories. When the Club of Rome asked for a global forecast based on new computer technology, Jay Forrester already had the model approach (system dynamics) and the technology at hand. However, this example of a new culture did not just thrive on the iterative capabilities of the digital electronic computer. The technology also shifted the standards regarding when a prediction counts as valid. With a slight oversimplification, iterative methodology trumps empirical precision in the context of complex systems. Overall, the findings on (co)evolution, variety, and structure provoke three questions that the remainder of this chapter addresses. The first two questions look back; the third looks ahead.

(1) Why does that prediction feature so prominently in science and engineering—and prove to be so rich a topic for historical and philosophical studies—yet is examined so rarely? Our tentative answer identifies an

unfortunate confluence of philosophical methodology and the alleged status of mathematics.

(2) What is the status of the classification into four cultures? Are there in-between cases that combine features of several cultures? Yes, there are. We have repeatedly observed and analyzed hybridizations between cultures of prediction. Furthermore, building such hybrids seems to be a key feature of engineering.

(3) What about history dragging on? Will new cultures of prediction evolve? In all likelihood, yes. We argue that the current hype of deep learning is part of a new culture of pure prediction.

9.1 Prediction—Functional Unit versus Coevolution

When the first question asks why the existing literature reflects so rarely on the importance of prediction, it is tempting to give the blunt reply: because prediction is of so little historical or philosophical interest. This statement must be qualified in two ways: First, prediction is receiving increasing attention in connection with recent machine-learning methods. We address this issue in our reflection on the third question. Second, the generally weak coverage can be attributed to a widespread but misleading view that prediction is a functional unit. Nobody questions that predictive capability is important, or even whether it is at the heart of scientific and engineering activity. But it is seen as exactly that—an activity for engineers and scientists. They learn how to apply existing mathematical models and techniques that yield predictions and also how to develop new ones. *That* math instruments yield predictions counts as almost trivial, whereas *how* they do it is seen as a concern of the specialist, neither interesting for nor accessible to the nonspecialist. From such a perspective, making predictions with mathematical means appears as a functional unit.

This unity is a myth that can continue to exist only as long as historical and philosophical investigations do not pay closer attention. Each of this book's chapters provides a fundamentally different picture. Prediction is growing out of a very dynamic interaction—not resulting from the execution of logic, data, and math but from the coevolution of epistemology, technology, mathematics, and their social organization. Gaining an understanding of this kind of coevolution requires the combination, and sometimes inventive mix, of disciplinary perspectives from philosophy, history, and science studies.

A number of insightful works covering different facets provide a crucial orientation here. Of course, the social turn is prominent since at least Thomas Kuhn's seminal work. We are, like Kuhn, indebted to Ludwik Fleck's studies of thought styles (Fleck 1979) that bring together (or encourage us to do so) philosophical, historical, and sociological perspectives for describing configurations.[1] More recent studies like those of Andrew Pickering on *The Mangle of Practice* (1995) underline how important it is to take into account several factors together and to integrate historical and philosophical with social and practice-oriented angles. However, a mangle exerts a lot of pressure to bring the mangled pieces into form. In our examinations of prediction, we found that such pressure does not play a leading role. Hence, we preferred the less precisely outlined, and therefore more open, terms of culture and mode. This framing should invite one to see prediction as the complex and multifaceted activity that it is—even and especially when it comes to mathematical tools.

There is also illuminating work that shares the particular focus on mathematization and mathematical tools. To be sure, many accounts of mathematization address the tensions and upheavals that have arisen as a result of mathematics' new role in methodology, in epistemology, and, more generally, in the way we view the world. Peter Dear (2006), Donald MacKenzie (2001), Michael Otte (1993), Ian Hacking (1990), and Lorraine Daston (1988) provide exemplars from which we have learned a great deal. Mark Wilson (2017) derides philosophies as "theory T thinking" when they assume mathematical relationships are logic-derived notions instead of the more variegated and partially incoherent strategies they are in practice. Such approaches "erase the very detail they require to resolve the conceptual questions before them" (57). For this reason, we designed our study of prediction in a way that should escape a similar criticism and provide a history and an epistemology of the use of mathematical tools for prediction and, in this way, establish it as a profitable object of research.

Admittedly, such an analysis initially produces plurality. After all, it is also directed against a unity viewpoint. In strong agreement with this, Joseph Pitt (2011) argues that doing philosophy of engineering and technology means being oriented "against the perennial" (viii). Paul Feyerabend (1999) also uses pluralism as a lever against an idealizing philosophy, and Nancy Cartwright (2020) demands from science studies that they value pluralism, particularity, and practice—when science works, it works with a "tangle." However, there is a path back from detail and concretion to general significance. Analyses of rich practical cases allow something like an "ascension to the concrete"—to use Marx's (1973) phrase targeting abstract philosophy—that is open to the unexpected, open to the strange turns in how prediction is achieved—or thought to be achieved—by mathematical means (101).

9.2 Hybrid Cultures

The second question targets the classification of cultures of prediction into four types—rational, empirical, iterative–numerical, and exploratory–iterative. Distinguishing rational from empirical sounds catchy and has proven helpful in many sorts of analyses. However, such distinctions resemble ideal types (à la Max Weber) that aim at structuring an investigation rather than fully capturing an actual configuration. In other words, these four types help us to gain insight into the history of prediction in which, nonetheless, hardly any individual case can be classified as being pure. No case is entirely rational, and none is entirely empirical. But this is not a weakness of the typology. On the contrary, it indicates applicability. Think of Isaiah Berlin's (1953) brilliant essay on persons being either hedgehogs—knowing one big thing very well—or foxes—that is, having some knowledge about many things. Although no person is completely a hedgehog or completely a fox, it is instructive to see them classified this way. In other words, the typology of cultures of prediction used in the previous chapters (rational, empirical, iterative–numerical, exploratory–iterative) should initially serve to identify relevant differences. At the same time, each chapter developed a richer and more adequate picture and identified hybrids between cultures. We make no claim to be exhaustive.

Not even exploration is restricted to the exploratory–iterative culture (presented in chapters 4, 7, and 8). Some elements of exploration occurred earlier, such as, for example, the robustness tests (Cole and Curnow 1973) of Forrester's world model (chapter 5). Building computer models always requires some element of exploration. A main reason for this is that model behavior—that is, the actual result of many iterations on a particular machine, using a particular algorithm and software, and so forth—must be sounded out in iterated runs. Exploration has a long trace already present in the mainframe culture. As computers became more available and more interactive, the role of exploration grew. In the 1970s, only a couple of pioneers were imagining that computers could strongly support an exploratory approach. One of them was J. C. R. Licklider et al. (1967) who imagined real-time feedback as direct interaction between researcher and computer:

> The modeler observes through the screen of an oscilloscope selected aspects of the model's behavior and adjusts the model's parameters . . . until its behavior satisfies his criteria. To anyone who has had the pleasure of close interaction with a good, fast, responsive analog simulation, a mathematical model consisting of mere pencil marks on paper is likely to seem a static, lifeless thing. (282, cited according to Waldrop [2001, 98])

Licklider was clear that his vision would not work because computers were too costly at the time. This economic factor changed only after the advent of microprocessors and their shrinking production costs.[2]

Furthermore, our investigations brought to the fore how hybrid cultures are of particular significance in engineering. Tartaglia pioneered a rational culture of prediction, and Galileo developed this culture further by giving mathematization a new significance. Robins, however, clearly pursued a very different culture in which he wanted to combine empirical with rational aspects of mathematization (chapter 2). His approach was controversial exactly because it mixed both ideal types and thus located prediction in a different coordinate system. Chapter 3 on the mathematization of mechanical engineering deals with a similar situation. The German Anti-Math movement at the end of the nineteenth century developed in stark opposition to the rational viewpoint of an earlier generation of engineers. However, proponents such as Robert Thurston at Cornell or Carl Bach at Stuttgart advocated a new conception of mathematization that was not concerned with eliminating rational elements but with combining them with empirical ones in a new way, thus establishing a hybrid culture of prediction in engineering.[3]

As new cultures of prediction develop, this diagnosis is reinforced by the technology of the computer. The iterative–numerical culture associated with centrally managed mainframe computers was distinctly different from the exploratory–iterative culture that emerged only with readily accessible computers. This expansion rested heavily (but not exclusively) on iteration. Certainly, iterative algorithms were in use for many centuries before anything such as a digital computer was invented. However, the ability to iterate so rapidly changed the conception of mathematization in fundamental ways, and the computer as an instrument allows, and also calls for, a different kind of social and cognitive organization—the effects of which have only gradually become (and are still becoming) apparent. Although situated very differently, the inquiries into computational chemistry, Bayesian statistics, and thermodynamics engineering reached a joint accord, namely the mutual significance of exploration and parameterization. They build a flexible link for use in practice and thus a boost for hybridization.

In fact, there is a good reason why the outcomes delivered by prediction are not pure types but mixtures: they are produced by the coevolution of mathematics, technology, and social organization. And this happens in an open way that is not bound to pure types. What prediction means, what is accepted and when, what is considered a promising methodology, and how this methodology is organized socially and cognitively—all this develops in mutual dependence. Coevolution proves to be seminal to our dynamic history of prediction in science and engineering. This consideration concludes our brief look back. We now turn our heads and look ahead. One could say that we bring to bear what we have learned about the history of prediction to understand the present of prediction.

9.3 Looking Ahead: A New Culture of Pure Prediction?

The third and last of our questions inquires whether a new culture of prediction might evolve.[4] Are we not witnessing the emergence of a new culture of prediction—a culture that involves machine learning and deep neural networks? A culture that is discussed widely in the media under various headings ranging from AI to digitization? A culture that might aptly be called one of "pure prediction"? In the remainder of this chapter, we shall reflect on the suggestion that the recent movement in AI and machine learning heralds a culture of pure prediction. We begin with a snippet of a story.

Automated driving is a recent example that draws much public—and commercial—attention and illustrates how problems of prediction coevolve with computational tools. When determining what the appropriate controls over an automated car will need to be in the next instance of time, a number of predictions have to be processed such as whether some object is going to move across the street, whether the brakes have to be activated because there is a stop sign, and so forth. The adequacy of predictions can be a matter of life and death. On May 7, 2016, a Tesla car on autopilot with a human driver not touching the steering wheel had a fatal accident on a Florida highway. Such cars have a number of sensors on board including cameras and radar. An eighteen-wheel truck started crossing from the median, with a big white trailer that was difficult for the camera to discriminate against the bright sky. However, radar should have easily detected the object. The Tesla crashed into the trailer without any attempt at braking. The autopilot system had made a wrong prediction. The trailer, according to Tesla's explanation in the subsequent investigation, had been taken to be an overhead sign—no breaking required. This accident was the first fatal accident with an automated car. All kinds of questions arise regarding such issues as regulation, safety, responsibility, and trust that require us to reflect on the predictive system, its properties, and its uses.

We would like to point out that the car was controlled by a machine-learning computer model—that is, an artificial neural network (ANN) that classified the incoming data (from cameras, radar, and other sensors) and delivered the predictions needed to operate the car. Because it was the first incident involving an automated car, the case attracted a great deal of public attention, and the obvious question was: If the ANN produced a tragically wrong prediction, why did it do so?

The answer is hard to give for various reasons. One of them is that the predictive system is proprietary, and Tesla has a vested interest in not providing real insight. Because this situation is typical for the use of software that is intended to produce a commercial profit, difficult questions of regulation and accountability arise. Another reason that makes the question so difficult to answer is the mathematical and computational tool itself—that is, the ANN.

ANNs classify according to a learning algorithm that is easy for a computer but impossible for a human being to follow because it works with a quite generic model of input–output behavior (stop sign in—brake activated out) that is extremely voluminous when it comes to mathematical operations with data. In particular, training these ANNs with the help of a great deal of data involves the adjustment of millions of parameters in the model.[5] Theoretically, these systems are extremely flexible and can emulate almost any input–output behavior. Based on sufficiently rich data and after adequate adjustment of the parameters, the predictions should be fine. Importantly, the adjustment, called "learning,"[6] happens automatically following an optimization algorithm.[7]

In the model, the trailer was apparently similar to a road sign—but why exactly this was so escapes human explanation—it is just what the ANN has learned in its machine way. Just imagine a photo portrait of ours. Most people we know (including ourselves) can easily recognize it when displayed on a cell phone. However, the computer (of the cell phone) operates with an array of pixels, and each frame contains about a million of them. If we were to receive the information about each pixel value in a very long series, we would have no idea what the image displays. However, this is what machine learning with ANNs achieves: the identification (of a large set) of correlations between (large) sets of pixels. These correlations are what distinguishes our portrait from other pictures (from the perspective of the ANN). The question of why this image differs from others—or better: why images of us differ from images of other persons or objects—is pointless. It is simply a different set of correlations. In other words, the question as to how a model can adequately represent a target system has become obsolete.[8]

Being able to work with the statistics of correlations is a feature of deep learning or AI working with multilayered ANNs. This feature rests on the enormous iterative capabilities utilized in data input, parameter adjustments, and optimization. Hence, this sort of prediction belongs to a culture that can be called iterative. However, there are a number of issues that lead us to wonder if a new culture of "pure prediction" is developing.

Therefore, the output of such ANNs leaves little room for reasoning about what exactly caused it. Of course, sometimes concise answers do exist such as when a relevant type of input has not been in the training data. Because the network did not see this kind of data in the training phase, it is no wonder when it does not react adequately. Normally, however, there is not much more to say than that the data and the learning together produced the prediction (as was the case in the Tesla story). Everything that makes a difference is contained in the subtle differences of adjustable parameters. However, there are millions of them, so their relationships are hard to oversee. Compared to human powers, ANNs are statistical monsters. They make predictions possible even in the most intricate situations but do not allow any queries about reasons. In short, ANNs produce predictions but not much else. So, maybe pure prediction looks like an adequate label.

Do ANNs belong to one of the computer-related cultures of prediction that we examined: The iterative–numerical or the exploratory–iterative one? First of all, ANNs belong there because they are certainly characterized by iteration. However, they also have unique features that do not coincide readily with either of the two cultures. Adjustable parameters have been a hallmark of the exploratory–iterative culture. Our case studies on computational chemistry (chapter 4), Bayesian statistics (chapter 7), and engineering thermodynamics (chapter 8) introduced adjustable parameters as an important supplement and extension to theoretical structure. This extension created the malleability necessary for prediction in practical situations. Hence, theoretical structure was combined with an empirical component in a way feasible only in an exploratory and iterative manner. ANNs take adjustable parameters to the extremes. Their successes seem to depend on the capability of adapting millions of parameters in a systematic way. Moreover, because the model structure is generic, the prediction seems to depend *solely* on these parameter values. This minimizes the theoretical component; or, more precisely, it shifts the theoretical component from a target domain (no mastery of chess or grammar necessary on the side of the modeler) to mastery of the computational instrument.

Another unique feature concerns the infrastructure. Data such as comprehensive image inventories from the internet are usually not stored locally.[9] Often, the actual optimization is also outsourced, typically to a software suite such as TensorFlow that runs on a platform maintained by Google. Thus, the exploratory–iterative mode of prediction has been adopted by a new centralized infrastructure. Although it is centralized as in the mainframe regime, it is readily available (or those parts of it are that some company thinks in its interest to make available). Moreover, the exploratory part is automated; it consists in adjusting the parameters almost entirely independently from the modelers, thus contributing to opacity.

The difference in infrastructure is closely related to a difference in social organization. One highly visible feature is that there is a host of competitions set up to achieve a given predictive task to the best degree or with the lowest failure rate (as on the platform Kaggle). Such competitions attract attention from various groups and have established an arena independent of academia (notwithstanding the fact that typical participants have had contact with universities). When data and software are provided on the internet, participants can act independently from resources provided by a university or other academic institution. These competitions function as a market from which big companies recruit scientists and programmers.

Importantly, the methodology together with the infrastructure create a new situation when it comes to policy and regulation. The quality of predictions depends on the quality of the (training) data. Because the quality of data is (still) ill defined, main actors take the quantity of data as a proxy. Today, data such as those that Tesla collects while developing its automated car count as a commercial treasure (not to mention Facebook and other actors in the field). Whereas the collected data are proprietary, government interventions such as regulating when a car has to apply its brakes depend on access to these data. And therefore, practice is heading for a conflict as far as regulatory measures—or better, their justifiability—is concerned.

In an epistemological respect, ANNs are characterized by opacity. In philosophy, epistemic opacity is discussed as a feature of computer simulation (Humphreys 2009; Lenhard 2019). The case of ANNs is special. They emulate all kinds of relationships through an extremely high number of statistical correlations. A face or an image is a correlation pattern of pixels. And while humans are used to operating with all kinds of images, they are at a loss when it comes to patterns. Opacity is an obvious obstacle to explanation. Not very astonishingly, and in response to the successes of ANNs, there is a recent call to develop "explainable AI." To the extent that opacity is a central feature of ANNs, seeking a strategy of explanation seems like trying to climb a ladder that has no rungs.

From a historical and philosophical perspective, prediction challenges the search for an explanation. This tension has been a constant companion to the entire discussion about explanation since the beginning of modernity—or actually even longer: ever since mathematics played any role whatsoever in considerations of epistemology and practice. A basic viewpoint is that the ability to predict shows something important. What this important thing is, is something that different cultures are anything but unanimous about—as we have shown. Nevertheless, the following applies across the board: the quality of the prediction redeems a claim that itself is based on other properties than merely the fact of the prediction. In some way, whatever is able to give good predictions has got something right about the world, or about that fraction of the world under investigation. And this something is the fundament and the true source of the predictive capability.[10]

Remarkably, the new culture seems to turn this upside down: prediction happens on the basis of a method, or a generic model, whose representational properties are in question or even inaccessible. We have seen that this is a trait of computer-based approaches in all of the iterative cultures. This kind of problem is relatively new to mathematics. It defies attempts to use mathematical tools exactly to avoid problems of opacity.[11] The principled stance toward the predictive machinery is then a bit like that toward the predictions of an oracle. However, these problems are not new to many parts of society in which division of labor makes people rely on the expertise of others. In such circumstances, this potentially new culture of prediction has not only an epistemic and social character but also a decisively political one.[12]

Practical work with ANNs is in flux, with epistemological, mathematical, and political questions arising—sometimes interrelated with each other. Overall, says the historian's caution, it is still too early to judge whether a new culture is emerging. Thus, even though our study of prediction does not provide an answer to the third question, it does provide a benefit. It provides a rich historical background and a flexible conceptual tool kit to reflect on the ongoing dynamic of prediction in an illuminating way.

Notes

Chapter 1

1. A few recent historical analyses, including Lynda Walsh (*Scientists as Prophets*, 2013) and Jamie Pietruska (*Prediction and Uncertainty in Modern America*, 2017), detail how making predictions has been related to manufacturing certainty. Arthur C. Clarke (1973, chapters 1 and 2 on *Hazards of Prophecy*) gathers stories in instructive detail in which prophets, including scientific ones, have failed. Walter Friedman (2014), for instance, tells "The Story of America's First Economic Forecasters" who, like Roger Babson, were economically successful without following any scientific methodology; or who, like Irving Fisher, advocated advanced mathematical methods but failed spectacularly to predict the Great Depression even when it was already underway.

2. As examples in which the philosophy of science addresses prediction, see the accounts by Peter Achinstein (1994) and Stephen Brush (1994) who discuss the relevance of predicting new phenomena versus explaining known phenomena.

3. Often, these books link prediction with a new computer-driven way of processing data. Examples are Nate Silver (2012); Viktor Mayer-Schönberger and Kenneth Cukier (2013); David Orrell (2007); or Eric Siegel (2013). Others throw light on the—important though also dubious—role predictions play in the economy. Examples are Nassim Taleb (2007) and William Sherden (1997); or in the broader society, Halpern (2000).

4. Compare the edited volumes by Lisa Gitelman on *Raw Data* (2013), which examines the interlinked history of data processing and mathematization; by Dan Sarewitz et al. on *Prediction* (2000), which focuses on scientific predictions in policymaking; or by Matthias Heymann et al. on *Cultures of Prediction* in meteorology (2017).

5. The works of Thomas Hughes elucidate the system character of technology far beyond the role of the computer. Importantly, the notion of culture, like the notion of the system, entails a certain interaction of components that reinforce the system or culture. This topic has been explored by Hughes (2004) and Levin (2004).

6. This literature began to develop about two decades ago with Paul Edwards's *The Closed World* (1996) and continued with David Mindell's *Between Human and Machine* (2002), Jon Agar's *Government Machine* (2003), Atsushi Akera's *Calculating a Natural World* (2007), and Fred Turner's *From Counterculture to Cyberculture* (2006). One decade ago, it included Paul Edwards's *A Vast Machine* (2010), Joe November's *Biomedical Computing* (2012), and Ron Kline's *The Cybernetics Moment* (2015).

7. Cybernetics is a prominent field in which prediction and control nearly merge—see Kline (2018), who argues about mathematical models in particular. Without sharing the focus on mathematical means, Miriam Levin's edited volume *Cultures of Control* (2004) exemplifies how technologies and ideologies stabilize into "cultures." She adopts a broad historical perspective and takes the culture of the Enlightenment as the quintessential culture of control.

8. Peter Dear's book on *The Mathematical Way in the Scientific Revolution* (1995) or Peter Machamer's study on Galileo, mathematics, and a new world view (1998) are but two out of a truly impressive body of scholarship on mathematization in the seventeenth century. Many aspects of the history of probabilistic and statistical reasoning are well charted by a group of interconnected works including the two-volume edition of Lorenz Krüger et al. (1987) on *The Probabilistic Revolution*, Lorraine Daston's *Classical Probability in the Enlightenment* (1988), Ian Hacking's *Taming of Chance* (1990), and Ted Porter's *Trust in Numbers* (1995).

9. William Aspray's edited volume on *Computing before Computers* (1990a) is one of the relatively rare contributions that looks at traditions and methods that bridge pre- and postdigital computer methods of computation. The edited volume by Lenhard and Martin Carrier *Mathematics as a Tool* (2017) provides another example. It contains a chapter by Ann Johnson that explores the idea of discerning "Rational and Empirical Cultures of Prediction" (2017) in the ways engineers use mathematics. Andrew Warwick (1995) argues that the main feature of mathematical predictions is their exactitude rather than their certainty. A particularly useful account is Peter Dear's *The Intelligibility of Nature* (2006), in which he explores a long-term tension between two positions: one that sees science as natural philosophy linked to knowing and the other as instrumentality linked to doing. These positions are ideal types, not fully and purely realized at any one time but effective in guiding development—much as the modes of prediction we discern.

10. We do not detail either the practice turn or the somewhat related model turn here. We just point out that recent accounts of computer modeling and simulation make good use of looking at practices (see Humphreys 2004; Lenhard 2019; Morrison 2015; or Winsberg 2010).

11. Donald MacKenzie has contributed an entire series of books on this. *Mechanizing Proof* (2001) and *Engine, Not Camera* (2006) are close to our study in the way they situate computational tools in a longer history of mathematization. We also learned much from Michael Mahoney's *Histories of Computing* (2011), which is concerned

with the ways actual scientific problems and questions are reparsed for computational approaches.

12. On this point, Andrew Pickering elaborated the metaphor of *The Mangle of Practice* (1995), Bruno Latour (1987) speaks of a network that connects technological and human actors, whereas Akio Akera argues for an "ecology" (2007). All three are innovative attempts at capturing mutual dependence along a process of coevolution that involves technologies, methods, concepts, and institutions.

13. We avoid most of the well-known subject matters of mathematization such as astronomical calculations or statistical mechanics, focusing instead on cases that are new to the literature such as the struggle over engineering mathematics in the late nineteenth century or the development of computational fluid mechanics.

14. This diagnosis is very similar to that of Lorraine Daston and Peter Galison in their illuminating book *Objectivity* (2007). They note that forms of objectivity are not displaced but that they are joined throughout history by further forms of objectivity. We share the emphasis on instrumentation, epistemology, and how they interrelate. Both objectivity and prediction tend to be seen as monolithic notions—but the two studies each show that this viewpoint does not hold water when practices are scrutinized. An important difference is that we pay more attention to hybrid and immature variants. Furthermore, in our analysis, the tools studied (mathematical tools) are located in an extraordinary systematic context—making the results all the more surprising.

15. We do not differentiate between prediction, forecast, and other terms of similar meaning. Some authors intend to fix the terminology—for example, Erich Jantsch who declares that a forecast is a probabilistic statement whereas "a prediction is an apodictic (non-probabilistic) statement, on an absolute confidence level, about the future" (Jantsch 1967, 15). Friedman (2014, x) differentiates forecast and prediction in the following way: "To forecast is to make a prediction using tools not easily employed by the general public but requiring expertise." And he points out that forecasting has an ambivalent meaning as both predicting the future and shaping (casting) it (xi). We share his point about shaping; and this book details how methods, tools, practices, and goals of prediction (forecasts) coevolve. However, because there is no standard terminology, we use prediction and forecast interchangeably and indicate when absolute confidence or other properties are in play.

16. We are well aware that this is not a precise description of technology. We use the notion to refer to computers that are easily and cheaply available to researchers, a property that machines such as labscale minicomputers, workstations, or PCs share to varying extents.

17. In a way, this mode is the computer-based new edition of the hybrid rational–empirical mode of mathematization discussed in chapter 3 in the context of engineering knowledge.

18. Broad and controversial discussions around representation in philosophy of science try to pin down what it is that must be right.

Chapter 2

1. The historical and philosophical literature is far too voluminous to even attempt a condensed overview here. Koyré (1957), for instance, famously made a case for the rational (Platonic) roots of modern science. Amsterdamski (1975) is an example of the episteme-plus-techne viewpoint, whereas Shapin and Shaffer (1985) analyzed the (decidedly nonmathematical) experimentalism à la Boyle. This literature is so prominent that we can skip an exposition. Dear's *Discipline and Experience* (1995) is a particularly pertinent contribution in our context. He stressed how "the mathematical way" and its complexities codeveloped with the notion of experience, whereas we target the heterogeneity in mathematics-related approaches to prediction.

2. In an illuminating study, Peter Dear (2006) argues that natural philosophy and instrumentality are two ideal types to which science belongs simultaneously (7).

3. The term "mixed mathematics" reflects this practice. It has been used to signal the difference to geometry that counted as exemplary for homogeneous and stringent organization. In the examples we discuss, the nature of the mixture is controversial. The term itself, however, occurs later in the chapter in a historical context.

4. There are many historical accounts of various episodes but only one book about the development of ballistic science. That book, *Their Arrows Will Darken the Sun*, by Mark Denny (2011), is more of a popular account of the physics of projectile motion than a rigorous history of science or mathematics.

5. Because he was looking for a balance, it is apt to speak of a rational–empirical mode. We stick to "empirical" to underline the contrast.

6. *Nova Scientia Inventa da Nicolo Tartalea* (1537). A later edition, entitled *La Nova Scientia*, was printed in Venice in 1554; see Cuomo (1997) and Valleriani (2013) for a historical account and the latter also for a careful translation.

7. Just to avoid misunderstandings, we are using "analysis" in a generic sense here. Calculus and even Descartes are still waiting to be born.

8. Some historians see Tartaglia's entire work as motivated by the search for patronage (cf. Biagioli 1989; Cuomo 1998).

9. Tartaglia here is in line with the general Aristotelean rationale of starting with particular knowledge from experience and then producing general knowledge through reasoning. He was well-versed in Greek mathematics, publishing Italian translations of Euclid and Archimedes in 1543.

10. A consistent trait of science and scientific expertise is identifying (and redefining) what is reason and what is hazard. A culture of prediction revolves around a shared concept of how reasonable predictions are to be made.

11. This is the main point of Ekholm's (2010) valuable work on Tartaglia.

12. We owe this observation to Arend's comprehensive study (Arend 1998, 191–192).

13. Tartaglia knew this work well. In 1543, he was to publish the *Elements* in Italian, the first translation into a modern language.

14. Arend (1998) provides a well-argued example.

15. Sebastian Münster (1551) and Daniel Santbech (1561) also developed mathematical accounts of projectile trajectories. However, they did not envision Tartaglia's kind of innovative mixture but used standard (geometrical) approaches. Their accounts did not have much impact.

16. Tartaglia's amalgam also included "Archimedean reasoning" that is replaced in a later edition by "physical reasoning," which means including sense perception. Tartaglia never actually used data. The *Nova Scientia* mentions a one-time shot with a cannon to verify the 45-degree angle for maximum range, but the evidence remains unclear.

17. Seminal accounts include Alexandre Koyré (1957), who sees Galileo as Neo-Platonist rather than an experimenter, and Geymonat (1965) and McMullin (1967), who reclaim Galileo as an experimenter. McMullin sees Galileo as an advocate for the notion of freedom of expression being essential to the generation of useful science (reading Galileo's letters to Duchess Christina). Valleriani (2010) examines Galileo the engineer. Machamer (1998) gives a good overview of the "Galileo industry," including further literature.

18. On this tradition of Italian scholars of the sixteenth to the seventeenth centuries, see Drake and Drabkin (1969).

19. Cf. Biagioli (1993) and Westfall (1985).

20. Machamer (1978) and Lennox (1986) forcefully argue that this is the adequate category for defining Galileo.

21. Galileo (1974). Valleriani's (2010) findings indicate that the strength of materials would provide another case study for our claim about mathematization and conflicting modes of prediction (see also Johnson 2017).

22. Galileo was not a first-rank mathematician, and he was no pioneer at all concerning the role of algebra, sometimes considered the key to mathematization. His student Torricelli was able to achieve more exact results because he invoked more algebraic means. The much later "rigor movement" in the nineteenth century tried

to rely on arithmetical reasoning because the perspective had changed completely by then, and geometrical reasoning was viewed as not rigorous enough.

23. That an angle–range formula would be useful for gunners only if sufficiently simple is another issue.

24. Only calculus, elaborated by Newton, Leibniz, the Bernoullis, and others, would enlarge the mathematical toolbox.

25. See Erlichson (1998) for a reconstruction of how Galileo obtained his mathematical results.

26. The "statement of others" probably refers to Tartaglia who also stated the 45-degree angle but did not give the kind of mathematical derivation for it.

27. We rely heavily on Segre's work on Torricelli (1983, 1991).

28. They traveled well below sonic speed.

29. According to what has been called "Galilean idealization" (McMullin 1967), strictly speaking, wrong assumptions might be quite adequate to model physical processes.

30. It is unclear what exactly Renieri tested. He reported on gun shots at different angles. But Segre recomputed these and found the reported results inconclusive.

31. We do not fully agree with Michael Segre (1983) who argues that Torricelli's view in this work is that "mathematics does not describe reality" (489).

32. B. S. Hall (1997) analyzes reasons for the unpredictability of spherical cannonballs shot from smoothbore barrels, especially the incalculable spin placed on cannon balls by their final point of contact with the barrel (202, fn 76).

33. Dear (1995, ch. 6) highlights the mixture of demonstrative mathematical reasoning and causal physical reasoning. For a historical account, see Brown (1991).

34. Brett Steele has worked extensively on the case we are analyzing here (1994). In his contribution to the collection he edited on Enlightenment war and science, *Heirs of Archimedes*, Steele argues that looking at eighteenth-century ballistics as a failure, as many historians have done, is a mistake. Whereas we agree that the alternative between Plato and Aristotle does not provide an adequate contrast as a backdrop to analyze mathematization and modes of prediction, we also maintain that "a symbiosis (. . .) between the supply of mechanistic science and the demand for military capability" (Steele and Dorland 2005, 15) does not serve either.

35. Johnson (1992) gives a valuable account of Robins's biography but does not address our topic of mathematical tools and mathematization.

36. The observation can count as the first anticipation of the *sound barrier*. Today, the coefficient for air resistance is not seen as a constant but as a function of the

Reynolds number. The science and mathematics become intricate here. Denny (2011) and Long and Weiss (1999) provide accessible accounts.

37. This famous story is told in part by I. Bernard Cohen in his guide to Newton's Principia, see Newton (1999, 168–171) and by A. Rupert Hall (2009, 140, 152–156).

38. Truesdell (1984) includes an informative biographical essay.

39. Historian of mathematics Henk Bos (1980) welcomes using rational mechanics as a lens. He applauds Truesdell for giving a structured account of eighteenth-century mathematics. In most other accounts, according to Bos, the eighteenth century looks like a mere bridge between the seventeenth and the nineteenth. Truesdell is right in shifting the focus to rational mechanics.

40. Alder is critical of Steele for painting a too heroic and not sufficiently rich societal picture (1997, 91–92), whereas Bos raises doubts as to whether "insights gained through mathematical theory effectively influenced practice before the nineteenth century" (1980, 354).

41. Philosopher Mark Wilson (2006) has pointed out that numerical approximations might induce their own set of conditions different from (and alien to) what rational mechanics assumes. In Euler's case, his iterative algorithm may lead to false results because it assumes a Lipschitz condition that is not fulfilled at the highest point of the trajectory. How good the approximation actually is must be determined by empirical measurements: "We engineer a thinner hold on its appropriate measures of correctness than classicism presumes" (Wilson 2006, 176). We argue in chapter 4 that questions of numerical methods and computations lead to a new mode of prediction.

42. Both modes can claim lasting success. Robins's book went through several editions in English (while Euler's commentary was not translated into English until 1777) and was used at the Royal Military Academy (Sandhurst). Euler's method was in use up to World War II for low-speed projectiles (McShane et al. 1953, 305).

43. This frontier is still moving. Numerical methods together with electronic computing technology can answer many questions that are intractable with analytical methods, whereas such new mathematical instrumentation is opening up new questions (see Sengupta and Tatta [2004] on the Robins–Magnus effect).

44. Bashforth (1890) dedicates his introductory chapter to a review of measurement technology created in the later part of the nineteenth century (including his own electrical chronograph).

45. Gluchoff (2011) examines the American side and reports the work of Moulton at the Aberdeen Proving Ground.

46. Grier (2001; 2005) tells the story of the Aberdeen Proving Ground where a large number of people (mainly women) were employed as "computers."

47. Haigh, Priestley, and Rope (2016) tell this story in enlightening detail, revealing that many commonly held opinions are incorrect.

Chapter 3

1. Historians have highlighted different aspects of these changes. In his classical study, Monte Calvert (1967) scrutinizes the conflicts in the United States between established shop culture and upcoming school culture. In the process of professionalization, the turn to school culture is linked to the striving for high social status. According to historian Peter Lundgreen (1990, 33), social motives are the de facto drivers of professionalization, whereas the belief "that modern societies have characteristic needs that must be met" is a technocratic myth. Ed Layton (1986), in contrast, identifies industrialization as a driving factor in professionalization. Tom Hughes (1989) pushes the point one step further and defines the American nation by technology or rather by the system of technologies that includes social organization and education as components. Terry Reynolds (1991) provides a succinct historical overview of the traditions and sources that made up American engineering.

2. See also Böhme et al. (1978) on the scientification of technology. However, König (1993) weakens this claim by arguing that, in Germany, industry did not demand more scientifically educated engineers.

3. Gispen (1989) and Lundgreen (1990) give standard accounts in English. However, most contributions including Manegold (1971; 1980), Hensel et al. (1989), and Dienel (1993) are in German.

4. Archival material documented by Puchta (1998), formerly Hensel, requires a reassessment of the role ascribed to him in older literature.

5. Chapter 4 follows these tensions in computational chemistry over a large part of the twentieth century, whereas chapter 8 discusses them in the new context of engineering thermodynamics and computer simulation.

6. Recent literature is almost as divided as the actors have been. Layton (1971) is a classical on how the picture of engineering was modeled. Another example is Böhme et al. (1978), who argue that what happened was a scientification of technology. König (1993) counters this by pointing out that successes in industrial engineering often happened independently from science.

7. This might seem a mild irony of history for a staunch believer in theory such as Rowland. Maybe Hughes (1989) would feel confirmed in claiming the United States as a technological nation.

8. David Hounshell (1980) suggests that Rowland may have been prompted to strengthen his claims because of sour business dealings with Edison, the then paradigmatic inventor (i.e., nonpure) scientist.

9. Rowland also defended a strong normative and ethical priority, but we leave out that aspect (see Lucier [2012] for a summary).

10. In the late 1870s, he served as vice president of AAAS and, from 1880 to 1882, as the first president of the American Society of Mechanical Engineers.

11. Thurston acknowledged his forerunners: King's College, London; the University of Edinburgh; the Polytechnic at Zurich; and the Munich Laboratory, planned in 1871 by Linde. Linde also made a case for research and education.

12. Thurston (1896) made his first laboratory-based discovery in 1872 of "the exaltation of the normal elastic limits by strain" (270). The "normal" limits were those assumed on the basis of theory. Starting in 1875, Thurston acted as secretary of the "U.S. Board appointed to test Iron, Steel, and other Metals" and took over official testing tasks at Stevens from the government.

13. See Calvert (1967) on the stalemate between adherents of shop and school cultures that threatened to paralyze Cornell's engineering college. Because of his work as a consultant, Thurston was in good standing with professionals although he favored school culture (cf. Durand 1939).

14. Of course, the use of empirical data to modify design equations was not introduced by Thurston alone. Steinmetz, for example, used it for his theory of the induction motor (see Kline [1987; 1992] and the discussion of Steinmetz's law in section 3.5, this chapter).

15. For a discussion of applied science—discerning four usages of the term—and the autonomy of engineering knowledge, see Kline (1995). For a historical perspective on engineering education in the United States from Thurston onward, cf. Seely (1993).

16. Thurston consistently underlined the importance of good knowledge of higher mathematics for making engineering predictive (see also Thurston 1896, 280).

17. Thurston wrote again in 1893 to the president, Andrew D. White, confirming that the concept of a purely professional school that would make no attempt to provide a general education was highly efficient and successful (Calvert 1967, 102).

18. For instance, Thurston presented a two-hundred-page report on technical education to the American Society of Mechanical Engineers (ASME) in 1893 in which he elaborated on the structure of the Sibley curriculum.

19. Thurston (1893a; 1893b; 1893c), Burr (1893), and Swain (1893) all sang from the same hymn sheet.

20. Whoever visited Germany would have had ample opportunity to witness a (then) new Imperial style in buildings of all kinds that clearly expressed a political stance.

21. The historian Kees Gispen (1989) describes the particularly uncomfortable position of German engineers who struggled to become fully integrated into the (feudal) establishment. Gispen argues that the "rift separating *Technik* from *Bildung* and *Besitz* remained so wide and deep in Germany that engineers were forced to develop something like a counterculture and to compete rather than amalgamate with the dominant social order" (2). This is markedly different in France as well as in Britain (see Shinn 1980; Wiener 1981).

22. Adelheid Voskuhl (2016) details the role of both the history and the philosophy of technology in the engineers' desire for upward social mobility. She mentions the "social critic" Carl Julius von Bach who will serve as main actor in section 3.4 of the present chapter—although in his role as promoter of a new concept of mathematization.

23. Translating names of German institutes is notoriously difficult because not only did they change names but their English meanings changed as well. Polytechnical schools toward the end of the nineteenth century changed into *Technische Hochschulen*; and, later in the twentieth century, into technical universities. The formal name "university" once indicated that students were offered a full range of humanities. This relic of nineteenth-century Humboldtian universalism has more or less lost its influence on contemporary terminology. One more point on terminology: The German *Techniker* [technician] and *Ingenieur* [engineer] were practically synonymous at the time.

24. This history is covered mainly by German authors such as Hensel (1989, 1991) and Mauersberger (1980). On Reuleaux, see Ihmig (1989).

25. For instance, at Karlsruhe, maybe the institute providing the foremost theoretical education, Redtenbacher, who himself promoted a balance between theory and practice, had hired the mathematicians Clebsch (in 1858) and Schell (in 1861) who stood for this new abstract orientation. See Stäckel (1915) as well as Hensel (1989) and Otte (1989).

26. Reuleaux had an international reputation for proposing engineering as a science. Reuleaux, who had reported from the 1876 World Exhibition in Philadelphia about Germany's embarassingly poor reputation due to cheap and shoddy (*billig und schlecht*) products, was convinced that engineering's way out must lead toward mathematized science (1877). His theory of kinematics (1875, 1900) was built in a logical, axiomatic fashion. Although Thurston is reported to have admired him (see Durand 1939), in fact, Thurston followed a very different track on mathematization.

27. Also in 1893, the VDI had sent a collection of important research questions to Helmholtz at the physikalisch-technische Reichsanstalt. Helmholtz did what the engineers expected—he acknowledged the importance of the questions, but declared the Reichsanstalt to be unable to address the topics. This independently confirmed that engineers had to carry out experimental research in their own facilities. The VDI moved on to strike while the iron was hot.

28. It is the founding document of the "laboratory movement" (Ernst 1894).

29. Klein gave a series of lectures. The (originally English) publication bears the full title: *The Evanston Colloquium: Lectures on mathematics, delivered from Aug. 28 to Sept. 9, 1893, before members of the Congress of Mathematics held in connection with the World's Fair in Chicago, at Northwestern University, Evanston.*

30. The terminology does not seem to be fixed but oscillates between anti-mathematicians and anti-mathematics movement. We keep this vagueness by using the English title Anti-Math Movement.

31. The Anti-Math Movement is not well covered in English. Gispen (1989) and Lundgreen (1990) both cover this episode, but do not look into the role of mathematics in any detail.

32. Puchta (formerly Hensel) (1998, 191) retrieved letters between Bach and Peters, the acting secretary of VDI, that document the role played by Bach.

33. See, for instance, Dienel (1993), Gispen (1989).

34. Because Bach had read and discussed Klein's memorandum on the planned foundation to the state of Wuerttemberg's assembly of engineers (Klein 1896c, see also Klein 1896b) in early summer, he could make his mind up before things ramped up at the general assembly in late summer 1895. Klein, in turn, had probably also expected approval from the side of engineers, but found himself mired in a bitter controversy with some of them.

35. Puchta's (1998) findings show that the mathematicians had also consulted with Bach. Furthermore, he persuaded engineers such as von Lossow to tune down their originally much sharper opinion statements when they published in *VDI Nachrichten* (Bach 1899, von Lossow 1899).

36. One example is August Föppl (1897) who defended a standpoint similar to Bach's before the assembly of German mathematicians. Aurel Stodola (engineer at ETH Zurich), or Walter von Dyck (mathematician, later president TU Munich) provide related examples of activities (Stodola 1897; von Dyck 1898). Hashagen (2003) gives a comprehensive account of von Dyck's life and work. He reports that the Anti-Math Movement did not have much impact in Munich where a mathematized engineering science developed in mutual agreement between engineers (such as Linde, Foeppl, or Bauschinger) and mathematicians and physicists (such as von Dyck). Dienel (1993) also highlights the role of TH Munich against the background of the controversy being much less fierce there. Carl von Linde had established experimental orientation and a laboratory in the mid-1870s, something Thurston had already been watching. Dienel is right when he diagnoses that "fruitful symbiosis of theory and practice" (1993, 87) is the solution. But actors on both sides of the controversy agreed on this general level.

37. This process started in 1899, but it took a couple of years before all states had approved.

38. At the TH Munich, von Dyck (former student of Klein's, see Hashagen 2003) created a similar institute for "technical physics" in 1902.

39. Klein also saw a related challenge for educational concepts. Klein was convinced that instruction needed to be linked to application (see Menghini and Schubring 2016).

40. This part of the 1902 course appeared as volume III of *Elementary Mathematics from a Higher Standpoint* in 1928 and in a new English translation as Klein (2016).

41. Klein advocated the renewal of the (rational) French ideal ("*Wiederaufnahme des Pariser Ideals*") (1900a, 23/24).

42. The controversy between Riedler and Reuleaux is discussed extensively in König's (2014) biography of the two men.

43. Where the quote is in English and the reference id to a German text, the translation is our own.

44. Riedler criticized that mathematics education (for engineers) did not entail the determination of coefficients from data (Riedler 1895).

45. Otte (1989) stresses this point in his interpretation of Riedler's work.

46. This is a telling instance of a German word that translates into several English words that express the meaning much more precisely.

47. Riedler gives an insightful account of this viewpoint in his book *Emil Rathenau and the development of big industry* (1916).

48. Puchta (1998, 194) has documented that Bach had been an oft-sought counselor and a pivotal mediator and author in the Aachen resolutions. Also, the later activities by mathematicians took place in exchange with Bach. Puchta (1998) examines Bach's position on the role mathematics should play in the education of engineers. We complement this by looking toward research and examining the hybrid conception of mathematization.

49. Gispen (1989) lays out the precarious status of large parts of the engineering profession in Germany—despite the success of pioneers such as Bach.

50. This gave him first-hand experience with Franz Grashof and his rational–scientific approach to engineering. Bach's already considerable experience in engineering practice must have been a striking contrast to what he learned at Karlsruhe. Overcoming this contrast may well have motivated Bach to work out an alternative to the rational school.

51. There had been forerunners. In England, "Testing and Experimenting Works of David Kirkaldy" had opened in 1866. The first testing lab in Germany had

opened in 1868 in Munich (directed by Johann Bauschinger), and Carl von Linde had opened the first mechanical engineering laboratory, also in Munich, in 1871. Zurich, with Ludwig Tetmajer directing the "Festigkeitsprüfungsanstalt" since 1881, was an equally early example, see Ditchen (2016) for a history of materials testing laboratories in Europe.

52. We quote from the eighth edition 1920.

53. We quote from the seventh edition 1899. Mauersberger and Naumann (1998) appreciate how the book set standards.

54. Due to the impending termination of the monarchy, he was also the last one. At least, we could not identify another instance (outside the military).

55. Bach was experienced enough to direct his attention to those cases where accepted rules failed.

56. Maybe such criticism of Grashof and others brought Bach a reputation for being antitheoretical. However, such a reputation would be unwarranted, as we shall show below.

57. Reuleaux (1882/1889, 870), Grashof (1875, Vol. 1, 473–476), and Weisbach (1880, 1095–1096).

58. In more technical language, the common theoretical treatment assumes a constant elasticity module for each material, implying that stress and stretch are proportional.

59. The point of spring break, if the pun is allowed.

60. Bach's assistant Richard Baumann (1917) published a critical essay on the dubious origins of Hooke's law: equation (*) "does not present a natural law, but no more, no less than the simplest of all possible mathematical relationships" (118).

61. After additional experiments, Bach doubted whether there exists some law that would cover all materials (Bach 1898, footnote 1).

62. Sandra Mitchell (2009) discusses rules as pragmatic laws for complex situations. She focuses on the complexity of the natural world but does not touch upon engineering.

63. Philosopher of science Mark Wilson (2006) argues along similar lines when he points out that even the alleged homogeneous mathematical structure of rational mechanics resembles a façade of patches.

64. Maybe Grattan-Guinness (1993) is right in blaming the mathsphobia of historians of science for persistent blind spots. Nonetheless, we searched for accounts that entail some mathsphilia from the side of historians.

65. For the numbers, see Ditchen's (2016) history of materials testing institutes. Zielinski's (1995) study provides more details about Tetmajer as the founder of the EMPA.

66. For details on the accident, the gravest in the history of Switzerland, see Schneider and Masé (1970).

67. The beauty of these constructions was hotly debated at the time. The aesthetics of industrial building material can still be contemplated in Paris at the Eiffel Tower built in 1889.

68. Throughout this brief case study, we rely heavily on the excellent biography by Ronald Kline (1992). Kline (1987) discusses the theory of the induction motor and how physicists criticized Steinmetz's modifications of Maxwell's equations.

69. Low-hysteresis silicon steel was not available in the nineteenth century.

70. Taken together, the books authored by Hunt (1991), Nahin (1988), and Yavetz (1995) give an outstanding coverage of the historical, philosophical, and engineering–scientific aspects.

71. For more details, see Nahin who refers to the controversy as "the battle" (1988, 196).

72. August Otto Föppl (1854–1924), professor for technical mechanics and graphical statics at TH München 1893–1922, introduced Heaviside's vector calculus into Germany. In 1894, he wrote the first German textbook on Maxwell's theory. Quite similar to Bach and to Heaviside, he held that electrical engineering is theory-based, but not Cartesian—that is, rational (1897, 109)—and that mathematization is not directed toward foundational conceptions.

Chapter 4

1. Of course, many factors are at work in the complex history of QC. Excellent and multifaceted accounts of this history can be found in Gavroglu and Simões (2012) or Nye (1993).

2. See Ashford (1985) for more details on the biography of Richardson who ended up as a pacifist volunteer in a British Quaker ambulance unit of the French army.

3. Developed by V. and R. Bjerknes in the 1910s, this remained the leading approach up to the 1980s (Hunt 1998, xxvi). For more detail on Bjerknes and the history of weather prediction, see Friedman (1989) and Harper (2008).

4. The title of Richardson's first draft was *Weather Prediction by Arithmetical Finite Differences.*

5. Lynch (1993) reevaluates the claims about computing time—"the better part of six weeks"—that Richardson makes in chapter IX of his book and estimates that 6 weeks nonstop computing time are meant, so that a single person would need about half a year to finish if working 40 hours per week.

6. Richardson refers only to the work of mathematician Carl Runge who followed similar lines in numerical mathematics.

7. See the review of Richardson's book by Exner (1923).

8. Platzman was a meteorologist and a pioneer of quantitative forecasting. He was also part of the team that used the ENIAC in 1950 to produce the first numerical weather forecast.

9. Proponents of a computer-related culture of prediction have identified computational power as the critically scarce resource ever since.

10. Iteration has been acknowledged by Hasok Chang (2004) in his *Inventing Temperature* where he discusses "epistemic iteration" as a valuable strategy, indeed as a key to justification in a nonfoundationalist, coherentist setting (in particular, chapters 1 and 5). For Chang, epistemic iteration, is "most likely a process of creative evolution; in each step, the later stage is based on the earlier stage, but cannot be deduced from it in any straightforward sense" (2004, 46). He highlights the general significance of self-corrective processes in science, but is also clear that the concept of iteration is borrowed from mathematics.

11. Iterative approximation is a key notion in Newton's method (see Smith 2002).

12. The term "computer" came into being in the context of human—mostly female—workers calculating in an organized distributed way to produce, for example, ballistic tables (see Grier 2005). Thus, human computers were in fact organized much in line with Richardson's "fantasy" of computing the weather.

13. Babbage was unable to finish his projected analytical engine (see Hyman 1985).

14. Considerations on how to design iterative numerical algorithms that would help to solve problems in mathematics gave rise to numerical mathematics with Carl Runge as a forceful proponent (see Richenhagen 1985).

15. There are more options to examine the history of math-based predictions in chemistry. For instance, Evan Hepler-Smith (2018) studies the influential work of chemist E. J. Corey who formalized organic synthesis and built a computer program to assist (or replace) the intricate process of looking forward and looking backward in the design of a synthesis. Curiously, according to Hepler-Smith, chemists incorporated Corey's formal approach into their practice but largely without making use of the computer. However, we concentrate on QC because it features mathematization and prediction in an outstanding way.

16. For a typical proposal, see the definition of QC in Per-Olov Lövdin's "Program" that opened the newly founded *International Journal of Quantum Chemistry* in 1967.

17. After this second turn, QC is often referred to as computational quantum chemistry. Computational chemistry, a member of the "computational" disciplines that have

emerged multiply since the 1990s, transcends QC and denotes a wider array of quantitative modeling using computer-implemented techniques that includes handling databases, drug design, molecular syntheses, and nanotechnology simulations.

18. From the standpoint of the history of science, it is the early phase of QC that has been researched most thoroughly. Scholars such as Nye, Gavroglu, and Simões have laid out the field. The books by Nye (1993) and Gavroglu and Simões (2012) cover the early history up to the late 1960s. Both take the establishment of QC as an accepted subfield of chemistry as the endpoint of their narratives. We add a further twist by identifying a new turn in the 1990s.

19. Many historical and philosophical arguments hold that the acceptance of quantum mechanics does not imply reduction (see Gavroglu and Simões 1994; Harris 2008; Scerri 1994; Schweber 1990; Simões 2003, 2002). We shall add an argument against reduction from the combinatorial nature of computational QC later in this chapter.

20. Gavroglu and Simões (1994) contrast German and American traditions in research culture that run more or less parallel to the strands discerned here.

21. The use of artificial or fictional components has been picked up in the context of the philosophy of simulation (cf. Lenhard 2007; Winsberg 2003).

22. Hartree developed his iterative strategy for precomputer technology, but his approach became increasingly popular with the digital computer. For more historical detail as well as more quantum chemical context, see Park (2009). For a more general appreciation of Hartree's work, see Fischer (2003).

23. Hartree himself called the approach SCF before the method became standardly known as Hartree–Fock. The Russian physicist Vladimir Fock (1898–1974) had pointed out weaknesses in the original Hartree method.

24. This perspective ties in with discussions in philosophy of science about the role of models and especially "models as autonomous mediators" (Morrison 1999).

25. George Forsythe (1917–1972) provides another instance. He was a pivotal actor in the movement to turn computer science into an academic discipline. He built and directed the Stanford computer science department and promoted computers as instruments for mathematics that should become part of undergraduate programs (Forsythe 1959). November's (2020) entry in the *Complete Dictionary of Scientific Biography* gives information and further literature on Forsythe.

26. Michael S. Mahoney (2005) refers to the growing importance of the computational standpoint as "reparsing" the problem so that it can be solved advantageously by computer. Ann Johnson (2004) describes this privileging of numerical analysis motivated by computing instruments in "From Boeing to Berkeley: Civil Engineers, the Cold War and the Development of Finite Element Analysis."

27. Handy et al. (1996) give a historical sketch of Boys's work and the success of Gaussian functions.

28. Klaus Ruedenberg's work in the 1950s or Roothaan's (1951) widely cited paper are further instances of fairly systematic investigations of strategies of computational modeling.

29. For a consideration of conferences such as the one on Shelter Island that aspire to found new fields; see Schweber (1986).

30. Indeed, it was not mathematical rigor that was decisive, but rather feasibility within a given framework of technology and manpower. Or, to quote Paul Humphreys's (2004) motto for computational science: "speed matters."

31. The LCAO-MO-SCF label indicates a combination of computational strategies building on the work of Slater, Lennart-Jones, Mulliken, Hartree, and others. The amalgamated name reflected the pragmatic mixture of approaches combining their respective computational advantages.

32. Utilizing adjustable parameters is pivotal in engineering knowledge (see chapter 3 and also Vincenti [1990]). Chapter 8 discusses the intimate relationship between adjusting parameters and the exploratory–iterative mode of modeling.

33. It was initially called NCCC, the National Center of Computation in Chemistry.

34. Because the competencies and working conditions for modelers change over time, software has not lost any of its bottleneck-creating abilities today. Michael Mahoney makes a clear case for the importance of software in his "Software as Science—Science as Software" (2002) and his *Histories of Computing* (2005).

35. In 1962, the Quantum Chemistry Program Exchange (QCPE) was set up at Indiana University as a hub for the distribution of software. Initially funded by the military, it changed to an academic endeavor in 1966. The goals of QCPE were to collect and distribute software including basic verification—that is, checking whether the program runs, not whether models are valid. Bolcer and Hermann (1994, 33) estimate that roughly one-sixth of users were quantum chemists. For more details on QCPE, see NAS (1971). DENDRAL can be considered an early example of such a type of software: It was built to elucidate molecular structure from mass spectra and made double use of the computer. It was designed for automated experimentation to accumulate spectral data plus a computer program to interpret these data (see Lindsay, Feigenbaum, and Lederberg [1980] and the comprehensive study of the computerization of biology and medicine from November [2012]).

36. This mathematical term basically indicates the growth of computational complexity in relation to the size of the problem. $O(N^5)$, for instance, means roughly if the problem size is N, the number of computational steps grows like N^5.

37. However, Scerri (2004) argues that it is not clear exactly what characterizes ab initio methods in QC.

38. See, for example, Boys and Cook (1960).

39. Today, the term ab initio is still in wide use, although the two meanings are not adequately held apart. Engel and Dreizler (2011), for a typical recent example, distinguish the "ab-initio or first principles approach" based on the "true, fundamental Hamiltonian" from a model-based approach that studies a "suitable model Hamiltonian." They suggest that an ab initio approach would not be model-based. But such a view would be misleading because ab initio approaches all rely on computational modeling.

40. The recent discussion in philosophy of science about models convincingly supports this standpoint (see, e.g., Lenhard [2019]; Morgan and Morrison [1999]; Winsberg [2010]).

41. For instance, Pople and Segal (1966) introduced the so-called CNDO method, one of the most influential semiempirical approaches.

42. See, for example, the account in Krishnan et al. (1980).

43. Cf. Nye (1993) on "the hubris so characteristic of the quest for mathematical certainty" (261).

44. The sudden movement of DFT in the 1990s from the boundary of the discipline to the mainstream is part of a broader change from QC to computational QC. This claim is developed in Lenhard (2014). DFT is a shining example, but not the only one. Molecular dynamics (i.e., using continuum models for molecular interaction) arguably underwent a similar history.

45. The density itself is a function and functions of functions are often called functionals—hence the name "density functional."

46. Redner also documents the steady flow of citations from the physics community that set in more or less instantly after the publication.

47. Another important contribution came from the experimental side—namely, from new technologies in spectroscopy that provided very detailed data and thus the opportunity for refined parameter adjustments.

48. Weisberg (2013) contains an elaborated version of this account.

49. Regarding commercialization, Gaussian is not unique. The production of commercialized resources for computational chemistry accelerated in the early 1990s. Now, "customers had to buy commercial versions of MOPAC, AMPAC, MM3, Gaussian, and other popular programs to obtain the latest versions with the most features and most bugs fixed" (Lipkowitz and Boyd 2000, ix).

50. In addition, Gaussian prohibits scientists who develop competing programs from purchasing licenses for Gaussian. Whereas Gaussian claims this is standard practice, many chemists object. See http://www.bannedbygaussian.org, and Giles (2004) and Gaussian's response at http://www.gaussian.com/g_misc/silly.htm.

51. In recent work, Alexandre Hocquet and Frédéric Wieber are approaching this issue in the case of molecular dynamics (Hocquet and Wieber 2021; Wieber and Hocquet 2020).

52. Centralized (super)computers are still extant, of course. Networks and the internet have made it easier to gain access to computing centers. A significant fraction of supercomputer time was and is used for computational chemistry. The cultures of prediction are not strictly sequential. Different cultures of prediction coexist and arguably even intermingle.

Chapter 5

1. "Prediction," the volume by Sarewitz *et al.* (2000) concentrates on environmental sciences and includes a valuable essay by Naomi Oreskes (2000) on the related history of prediction.

2. Although mainframe computers were the dominant form only in the early decades of digital computing (1950s to 1970s), they still exist in many computing centers. There are also further cultures of prediction related to digital computers. Around the 1990s, easily accessible networked desktop computers added an exploratory component in a way that the mainframe prohibited, thus leading to an exploratory–iterative culture of prediction (examined in chapters 4, 7, and 8).

3. We shall restrict ourselves to naming just a small sample taken from the large body of excellent historical scholarship. Ceruzzi (2003) gives a standard history of computing, and Metropolis *et al.* (1980) gather perspectives from actors. In-depth studies of particular topics include those by Aspray (1990b) on John von Neumann, Agar (2003) on the bureaucratic practice of handling files, Grier (2005) on human computers predating electronic ones, and Mindell (2002) on the analogue branch of computing. Waldrop's (2001) more popular account tells a story that leads to the personal computer.

4. We are greatly indebted to studies of computer use that highlight some variant of coevolution, namely "ecology of knowledge" (Akera 2007), "trading zone" (Galison 1997), the "mangle of practice" (Pickering 1995), and also the system perspective pursued by Paul Edwards (1996; 2000).

5. Chapter 4 has already introduced these features of the mainframe culture of prediction.

6. Here are four particularly shining examples of this history. David Mindell (2000) locates the roots of the systems approach in the labs at Bell and MIT in which engineers developed radar. Donald MacKenzie (2000) examines the near failure of the SAGE system. Paul Edwards (2000) conducts an insightful study in which he relates Forrester's world dynamics model to simulation models of weather and climate. Edwards (2000) makes the point that a systems approach constructed "the world" as an object for science and science-based policy (222). And he sharp-sightedly observes that the encompassing nature of the models required a global system for obtaining the data to be fed into the models. Finally, Ron Kline (2018) reveals insightful contrasts between Herbert Simon, Stafford Beer, and Jay Forrester in regard to "the contested issues of prediction and control" (285).

7. He brought in the engineer Robert R. Everett as coleader.

8. Redmond and Smith tell the history of the Whirlwind project in great detail in a way close to the developers' perspective. Akera's (2007, chap. 5) analysis is more attentive to the problems that riddled Whirlwind.

9. SAGE stands for semiautomated ground environment, an antiaircraft radar system. It was one of the monumental military projects in the early Cold War that combined high ambitions with deep wells of funding. It is well covered by historical literature (see Hughes 1998, chap. 2; Hughes and Hughes 2000; Mindell 2002, chap. 8; and Edwards [1996, chap. 3], who stresses the political and institutional side, seeing Whirlwind and the SAGE system as an example of the typically "Closed World" of the Cold War). For a stance that explores the absurd valences of a large military system, see Stanley Kubrick's *Dr. Strangelove.*

10. A servomechanism translates electrical signals into mechanical movement. MIT was a leading institution in developing such equipment.

11. Captain D. S. Diehl, chief of the Aerodynamics and Hydrodynamics Branch of the Bureau of Aeronautics, in a memorandum (1944, cited in Redmond and Smith [1980, 7]).

12. John von Neumann, the exceptionally well-connected mathematician and science organizer who was working simultaneously at Princeton, Los Alamos, and some further locations, had written up the "first draft to the EDVAC." This document introduced the concept of the stored program computer—that is, control by software. The intellectual authorship of stored program computers is contested because von Neumann's "Draft" arose over joint discussions with Mauchly, Eckert, and others (see the multifaceted account of ENIAC by Haigh et al. 2016).

13. Akera addresses the Whirlwind project—in particular, the early exchanges between ONR, the Office of Naval Research, and MIT—under the heading of "Research and Rhetoric. Jay Forrester and Federal Sponsorship of Academic Research" (2007, chap. 5).

14. A large project is itself a kind of predictive machinery. Future costs and success must be predicted to secure funding. The many charts for progress and cost that Forrester presented were plausible to MIT and the navy when issued but later proved to be examples of predictive failures. In the face of exploding costs, one of his rhetorical strategies was to predict an even better product in the future.

15. On SAGE and the more general problem of reliability in complex computer systems, see the enlightening work by Donald MacKenzie (2000; 2004).

16. A list of R&D costs locates all early computers in the same ballpark of 0.5 to 0.7 million dollars (not value-adjusted). Examples are IAS: $650,000; Eckert-Mauchly UNIVAC: $400,000 to $500,000; ENIAC: $600,000: and Harvard Mark III: $695,000. The single exception was Whirlwind, which cost more than $3 million—that is, about as much as all other computers together.

17. The average repair time was four hours per day, and the cost for the tubes amounted to $32,000 USD per month (not value-adjusted), amounting to an additional 1 percent of the development costs per month.

18. See Waldrop (2001, 114) or, similarly, Redmond and Smith (1980, 206).

19. The mainframe culture still exists. Although current high-performance computers have left Whirlwind behind by several orders of magnitude, most of the software runs cannot be repeated very often. They have grown in pace with the machines and are still too demanding and too expensive.

20. This recalls the controversy between Klein and Riedler about engineering and the role of mathematics discussed in chapter 3.

21. Akera (2007) and Redmond and Smith (1980) give details on the memos and the correspondence.

22. Forrester learned the theory of feedback control and communication from the Servomechanisms Laboratory at MIT. Engineering was an important component in theories of information, control, and communication (cf. works by Bennett 1979; Kline 2015, 2018; or Wiener himself 1948). Lenhard (2019, chap. 6) discusses a contemporary controversy between Wiener (MIT) and von Neumann (IAS, Princeton) on what should be the task of computer simulation.

23. A fitting description of how much the mainframe paradigm shaped the culture of prediction comes from Kenneth Olsen: "Indeed . . . to most people in the 1950s, that was what computers *were*: big, impersonal oracles sitting off in air-conditioned rooms somewhere, crunching data for big, impersonal institutions" (cited according to Waldrop [2001, 142], emphasis in original). Olsen was Forrester's student, then a leader in the advanced computing laboratory of MIT's Lincoln Laboratories, before he founded the Digital Equipment Corporation. Wes Clark, who pioneered standalone computers in 1961 with the Linc, expressed a similar viewpoint: "The only

surviving computing system paradigm seen by MIT students and faculty was that of a very large International Business Machine in a tightly sealed Computation Center: the computer not as a *tool*, but as a *demigod*" (Clark 1988, 353; emphasis in original).

24. In her book *The Future of the World* (2018), Jenny Andersson provides a historical and political analysis of futurism in all desirable detail. Futurism's prehistory lies in planning, national accounting, and statistics. For work on these earlier predictive endeavors, see Daston (1988), Desrosières (2000), Morgan (2012), or Porter (1995). Andersson identifies the decade 1964–1973 as the high point of future research. A good entry point to this research is the compilation by Alvin Toffler (1972). It was not only a bestseller in its time but lists about fifty books on futurism in the appendix—a broad outlook on what the future used to be according to futurism.

25. At the time, population growth and limited resources were being discussed widely. The Club entertained a Malthusian perspective—that is, it argued with a mathematical model based on the growth rate of the population and its consumption. Hardin (1972), for another contemporary example, gives a Malthusian-minded exposition of the "Tragedy of the Commons."

26. The term systems thinking denotes a larger movement to quantify the social sciences comprising cybernetics, system dynamics, systems analysis, systems engineering, operations research, game theory, and general systems theory. This is a well-researched area (see Heyck 2015; Kline 2015; Mirowski 2002).

27. The study by Meadows et al. acknowledges Forrester's claim; see William Watts's foreword to Meadows et al. (1972).

28. Actually, the model "world 3" in *LtG* is an elaboration of Forrester's "world 2" that adds detail while maintaining the same structure.

29. Chapter 4 observes this property in cases as different as the weather and the behavior of electrons and analyzes how iteration opens up a computer-based path to prediction.

30. Nonetheless, access is still highly restricted and regulated by doorkeepers like Forrester—a key feature of the mainframe culture.

31. Simmons (1973) draws a parallel to the technocracy movement of the 1930s. Engineers know the problems and their possible solution quite independently from being experts in respective fields. For instance, as Simmons observes, in his urban dynamics book, Forrester references only Forrester.

32. Lilienfeld (1978, 238) critically remarks that the system dynamics perspective drives abstraction to the point of meaninglessness.

33. Forrester openly agreed with Malthus but simply claimed that he himself could provide a more complete picture (see Forrester 1971, 2). See also Freeman's commentary on *LtG*: "Malthus with a Computer" (Freeman 1973).

34. This observation applies beyond the realm of science. For instance, predictions obtained from the oracle of Delphi were powerful because asking the oracle was an accepted go-to method.

35. Forrester insists on many occasions that computer modeling is inescapable because there is no other way to bring out counterintuitive characteristics of complex systems. However, this instantiates the tightrope walk: How are counterintuitive dynamic characteristics compatible with foreseeable system behavior?

36. Solow's growth model is an equation with just three variables: capital stock K, labor L (assumed proportional to population), and aggregate product or national income Y.

37. An algorithm is intractable when its execution takes too many resources to be practical. Obviously, tractability depends on available instrumentation. Adding the prices of all items I had for dinner is tractable for any skilled waiter with paper and pencil. Adding a list of a million prices is definitely not, although, in principle, it is the iteration of the same simple operation. However, it is perfectly tractable for a computer—in almost no time.

38. Another piece of evidence is the foundation of the World Future Studies Federation in 1973 that was heavily influenced by *LtG* (cf. Seefried [2015, chap. VII]).

39. Richard Ashley (1983) reviews important members of the second generation, including Bremer (1977), Guetzkow and Valadez (1981), Hughes (1980), Leontief et al. (1977), and Meadows et al. (1982).

40. A standard critique of the *LtG* as "Models of Doom" is Cole et al. (1973); see also Maddox (1972). Bloomfield (1986) offers a look back on the entire debate around *LtG*.

41. Interestingly, the critical evaluation by Cole and Curnow (1973) found that the robustness assumption is wrong. They found parameters indicating that the model behavior still fitted the period 1900–1970, but avoided the catastrophe, mainly by directing a higher fraction of capital into food production and raising the rate of progress from 1 percent to 2 percent.

42. In fact, the only real-world data that went into Forrester's study were about population size. The *LtG* took in more actual data.

43. See also Forrester (1971, ix, or 1969, principle 3.2–1) on model validity. *LtG* followed a similar line of argument (see Meadows et al. 1972, 20–21).

44. When considering the two camps, we follow Jenny Andersson's work (2012; 2018) on futurism. She identifies the first camp as "futurology" and contrasts it with a more utopian futurism. For the latter, she provides more examples such as Jungk and Galtung who founded "Mankind 2000" in 1968 with the purpose "to free the future from futurology" (Andersson 2012, 1423). However, Hartmann and Vogel

(2010, 16), who write about futurism in Germany, see a utopian futurism not in opposition to but as the forerunner to the Club of Rome study and the later environmental movement. Seefried (2015) stresses that the future movement changed the attitude from enthusiasm to skepticism about the future. She shows how different approaches lead to significantly different pictures of the future, so that one should speak of "futures" in the plural. However, her work does not reflect on the potential influences that the culture of prediction, including its technology and methodology, had on thinking about the future.

45. Doing historical research about different and changing conceptions of time and temporality has evolved into a prominent topic—see Hartog (2015) or Clark (2019) as examples. We follow Jenny Andersson's suggestion (2018) of juxtaposing Arendt and Koselleck.

46. Koselleck's "Futures Past" project (2004) aims to determine concepts of future held in the past.

Chapter 6

1. The text itself goes back to a manuscript that Ann Johnson presented several times and also to different audiences (historians, philosophers, computer scientists). Characteristically for her, these presentations were never identical but changed with their audience and with the progress made through discussions. Without doubt, her final version would have looked different, been more elaborated, and also been more sophisticated than the present text. I did not want to decide on its direction—and then an inevitably diminishing potential—but rather preferred to have this chapter documenting work in progress close to Ann's original writing, even if this is far from what she would have counted as ripe for publication. I am grateful to Robert Mullen and Michael Stöltzner for valuable discussions on CFD.

2. Readers interested in the history of computational methods might want to look at Ann Johnson's account of the Finite Element Method that developed and flourished in an engineering context (Johnson 2004). It is a fine example of readability and scholarship.

3. On model transfer and the interplay between maintaining identity and adapting to the situation, see the 2022/2023 topical issue of *Synthese,* edited by Chia-Hua Lin and Paul Humphreys.

4. Other candidates for comparison include Ursula Klein's "paper tools" (1999) and Andrew Warwick's "theoretical technology" (1992).

5. Kaiser's motivation for the title of his book *Drawing Theories Apart* was to contrast it with Latour's *Drawing Things Together* (1990).

6. Patrick Roache's (1982) book on CFD is an example.

7. Our historical approach is part of an emerging interest in code and software studies. This interest encompasses very different directions from work close to computer science—such as in Cynthia Rudin's Interpretable Machine Learning Lab (Duke University, North Carolina) or Katharina Zweig's Algorithm Accountability Lab (Kaiserslautern, Germany)—to work related to science studies—such as in Gabriele Gramelsberger's Computational Science Studies Lab (Aachen, Germany)—and to work related to political writing—such as in Mark Marino's Humanities and Critical Code Studies Lab (USC Dornsife, California). The obvious preference for the title "Lab" might perhaps indicate that the actors see themselves as making advances into yet uncharted territory.

8. Most of them make an appearance in chapter 2 in the context of ballistics.

9. See NACA report 1381, by Ames researchers Allen and Eggers (1958).

10. See the classic textbook by Liepmann and Roshko (1957), who mention the problem as one that cannot be predicted theoretically (5). It may be worth noting that shortly after this handover, Ames acquired a computer and eventually became NASA's high-end computing facility. See the Ames history by Bugos (2014).

11. His first name is alternatively spelled Francis. Harlow was the leader of the fluid dynamics group at Los Alamos from 1959 onward. At first, he worked on a room-sized IBM 701 mainframe computer (for more details, see Harlow's review [2004]).

12. The simultaneous development of finite element analysis shows that this strategy was common at the time (Johnson 2004).

13. Runchal argues in his recollection (2009) that Spalding created the practice of CFD as an engineering tool.

Chapter 7

1. The history of probability casts light on how science and society, in mutual interrelation, have developed ways to deal with uncertainties ranging from strategies in games to the prices of annuities. Whereas predicting the death of an individual might require divine foresight, estimating the death rate in a large population can get by with profane data. The historical and philosophical accounts by Daston (1988), Hacking (1990), and Porter (1995) cover the seventeenth to nineteenth centuries in admirably sophisticated ways.

2. The *Stanford Encyclopedia of Philosophy* has entries on the philosophy of statistics (Romeijn 2023) and a separate one on Bayesian epistemology (Talbott 2016). Taken together, these provide a guide to the large body of philosophical literature on Bayesianism.

3. This rule follows from the definition of conditional probabilities and is accepted unquestionably across all camps.

4. A sample of standard accounts from both sides of the Bayesian versus classical divide is Earman (1992), Howson and Urbach (1993), and Mayo and Spanos (2009). Hacking (2001) provides an accessible introduction. I openly admit that the picture I paint does not match the complexity of extant terminology in which subjectivism is pitted against objectivism, frequentist accounts of probability against subjective ones, and so forth.

5. Trying to keep things simple, I have glossed over internal differentiations of Bayesians. Neal's (1998) verdict that "there is (in theory) just one correct prior" might be controversial among adherents of Bayesianism. Corfield and Williamson (2001), for instance, discern subjective priors from objective (pluralist, logical, empirical) ones that, however, add further requirements for being rational while maintaining the principle.

6. On these differences, see, for instance, Lenhard (2006).

7. Bayes, for instance, had presented an example in which he did not know about priors and assumed equal distribution among possibilities. Neyman considered this step illegitimate.

8. Bayesians, in turn, would typically object that such agnosticism ignores relevant knowledge that actually *is* available.

9. McGrayne's book is about the eventual success of Bayesian approaches, so the quote does not reflect a bias against Bayesianism.

10. This figure resembles the one included in chapter 4 on the rise of density functional methods. Both depict a 1990s turn.

11. There are also some areas of statistical work that are closely connected to Bayesian methodology. One example would be causal analysis and Bayesian nets, a field following the lead of Judea Pearl (cf. Pearl 1995; or Williamson 2005). At present, it is a subfield of artificial intelligence and also philosophy of science. The flourishing of examples of this sort is not included in figure 7.1, which is already dramatic enough to motivate my analysis.

12. Zellner (1988) and Howson and Urbach (1993, 1st ed. 1989) argue for the superior rationality of Bayesianism independently of computational methods. Hence, they cannot account for the timeline of the upswing (except that it had to happen sometime).

13. Similar quotes abound. We picked this quote from McGrayne's book because of its atmospheric qualities. Here are two alternative quotes: "In fact, it may be argued that the main reason that the Bayesian approach to statistics has gained ground compared to classical (frequentist) statistics is that MCMC methods have provided the computational tool that makes the approach feasible in practice" (Häggström 2002, 47). The probability theorist here agrees with statisticians: "A principal reason

for the ongoing expansion in the Bayes and EB [empirical Bayes, jl] statistical presence is of course the corresponding expansion in readily-available computing power, and the simultaneous development in Markov chain Monte Carlo (MCMC) methods and software for harnessing it" (Carlin and Louis 2000, xiii).

14. For a short discussion on the form, beauty, and creation of Norway, see Adams (1979).

15. A third condition—that this should be carried out in your friend's apartment rather than your own—is of merely aesthetical concern and mathematically irrelevant.

16. The term "evaluation of integral" more aptly describes the point than "numerical integration."

17. We use the notion of location that suggests a geographical picture. Using the term state would be less intuitive but more apt to the generality of MCMC methods.

18. We greatly simplify matters in this discussion. Questions regarding how the space is defined or which technical conditions have to be satisfied are not important for the illustrative task we are pursuing here.

19. Persi Diaconis (2009), a leading expert on MCMC, describes the jaw-dropping surprise when he first saw how MCMC solved this task. R. I. G. Hughes (1999) gives a good account of the Ising model in the context of modeling and simulation.

20. Titterington (2004, 192) makes a case about a "Bayesian computational revolution"; Smith and Roberts (1993, 4) make a similar case.

21. MCMC is not necessarily Bayesian, but protagonists of Bayesian approaches were often developing MCMC methods to make Bayes's rule more relevant to practice.

22. Work on the Gibbs sampler started with Geman and Geman (1984) and gained popularity rapidly after the landmark paper of Gelfand and Smith (1990). The hit-and-run algorithm is related to the older Metropolis–Hastings and was proposed by Bélisle et al. (1993) and Chen and Schmeiser (1996). Other MCMC methods show similar timelines.

23. McGrayne (2011) tells the story in a vivid way on pp. 218–222. "The minute Gelfand saw the Gemans' paper, the pieces came together: Bayes, Gibbs sampling, Markov chain, and iterations" (221).

24. Philosophers of science have argued that computer simulation changes mathematics because simulation enlarges the realm of tractability (see, e.g., Humphreys [2004]). Regarding iteration, this is certainly correct. From this perspective, MCMC is a way to utilize the new tractability for problems of modeling.

25. We owe this formulation to an anonymous reviewer who helpfully inquired about senses of exploration.

26. This is very similar to the process of parameter adaptation of density functional theory; see chapter 4.

27. Consider the chapter by Datta and Sweeting (2005) on matching priors that are used to adjust priors *so that* the posterior distribution has the desired properties. Instances such as this abound—they are unavoidable when working in an iterative–exploratory mode.

28. The monographs by Levin et al. (2009) and Aldous and Fill (2002) document the settled state of the art; Diaconis (2013) provides an outlook on current progress and challenges.

29. Lunn et al. (2000) is the standard reference for WinBUGS.

30. See Lunn et al. (2009) and, for very brief historical remarks, the official website openbugs.net.

31. Other popular software includes JAGS (Just Another Gibbs Sampler), Stan (developed at Columbia University), MCMCpack, bayesm, or the SAS MCMC. Their differences are of no concern here.

32. To do justice to the full richness of Bayesian approaches, it would be necessary, as noted previously, to include approaches such as "objective" and "evidentialist" Bayes. Instead of entering a more detailed appraisal (see Romeijn [2023] and Talbott [2016]), this paper prefers a simplistic approach to Bayesianism and fully concentrates on the computational perspective.

33. Efron is famous for the bootstrap method that works in a frequentist guise. Therefore, he is presumably a significant witness in favor of Bayesian approaches. Moreover, he has repeatedly made claims that statistical inference has been transformed through computer use (explained in book length in Efron and Hastie [2016]).

34. Gill (2008) also favors "ecumenism."

35. Greenland further acknowledges that this theme is not new but also has been brought up repeatedly by Good (1983); Diaconis and Freedman (1986), which includes a discussion; and Samaniego and Reneau (1994).

36. This capability is based heavily on adaptable parameters, especially on priors that can be changed to increase the ability of a model to mimic the data—quite in line with our prior analysis of MCMC.

Chapter 8

This chapter was written jointly with Hans Hasse, head of the Laboratory of Engineering Thermodynamics at RPTU Kaiserslautern, Germany. This collaboration is itself a methodological experiment that provides access to scientific practice is an unusual way. And it was a highly enjoyable experience. As a consequence, the present chapter differs from the previous ones in that it pursues a systematic argumentation that is oriented more toward scientific and engineering practice and less toward history. This chapter is a revised and adapted version of Hasse and Lenhard (2017).

1. Rosen (2012) tells an engaging "Story of Steam, Industry and Invention" in great detail. In general, the history of the steam engine is so famous that we forego further references.

2. About twenty years after Clapeyron, Krönig, and Clausius derived the ideal gas law from kinetic theory.

3. A brief comment on terminology: Speaking about the "adjustment" of parameters invokes a field of similar terms with (only) slightly differing connotations. "Calibration," for instance, is used in the context of measuring instruments. Hence, talking about calibration of parameters makes models look a bit like precision instruments. "Tuning," on the other hand, has a slightly pejorative meaning, though it is used in some areas of science as the standard term. We chose "adjusting" because it seems to be neutral and does not appear to be a good or bad thing from the start—though this is not to claim that the present terminology is without alternative.

4. The claim about prediction and the role of adjustable parameters in the exploratory–iterative culture of prediction echoes, and thus vindicates, the findings in chapters 4 and 7.

5. Hasse and Lenhard (2017) therefore call adjustable parameters "boon and bane."

6. Although we cannot avoid discussing thermodynamic engineering, we shall do our best to avoid technicalities.

7. This statement mirrors our findings in chapter 4 almost verbatim.

8. The Web of Science has a system of disciplinary categorization. We made the following choice: if for one particular journal, one of the categories includes "engineering," the journal has ties to engineering. Because we limit our conclusions from bibliometric data to the crudest ones, such a simplistic choice is hopefully not invalidating.

9. The notion of models is a prominent issue in both history and philosophy of science; see de Chadarevian and Hopwood (2004), Lenhard (2019), Morgan and Morrison (1999), Morrison (2015), and Wise (2004), to name but a few references.

10. It coincides with other schemes of modeling such as R. I. G. Hughes's DDI (1997) account.

11. The notion of vapor pressure curve is already theory laden. We keep it simple and leave the background theory out of figure 1.

12. When addressing the intricate questions about correspondence and representation, we refer to Weisberg's (2013) work on a taxonomy of the relationships between model and target system.

13. Mark Bedau (1997; 2002) has highlighted that some properties can be known only by actually conducting the computational process of a simulation, and he has aptly called these "weakly emergent."

14. The two quantities that are compared do not need to be scalar quantities. They may have many entries or be, for example, trajectories over time. There are also many ways of carrying out the comparison.

15. The cooperation becomes even more entangled in those situations in which the measured quantities themselves are partly determined with the help of simulation or in cases in which the data used for the comparison come not only from classical experiments but also from other types of simulation.

16. Modifying the functional structure of the model is another option. Sometimes such modification can be interpreted as parameter adjustment (and vice versa).

17. See Oberkampf and Roy (2010, sect. 13.5.1) for a systematic proposal on how parameters influence the validation of simulations from an engineering perspective. Nonetheless, using mesh-size refinement can be a sound practice when an explicit estimation of errors is available.

18. Cf. Schappals et al. (2017) for the case of molecular simulation.

19. The issue of statistical (over)fitting is arguably an exception; see, for example, Kieseppä (1997).

20. There is a longstanding debate in philosophy of science about what ad hoc hypotheses are and what status they have (see, e.g., Grünbaum 1976; Leplin 1975).

Chapter 9

1. Cf. also Lenhard (2019) who applies the term styles of reasoning (referring to Hacking 1982) to shed light on computer simulation.

2. Whereas the story about the evolution of technology has been told, what we are highlighting is the coevolution of technology, epistemology, and institutional organization in the mainframe iterative–numerical culture of prediction.

3. One model for this kind of non-eliminative development is that proposed by Daston and Galison (2007) in their book on objectivity. They discern different modes of objectivity in which a new mode does not eliminate the older ones but adds a new possibility and enriches the picture.

4. Prediction has become a buzz word in recent reflections on machine learning, with Agrawal et al.'s *Prediction Machines* (2018) being just one instance.

5. The number of adjustable parameters is increasing rapidly in recent years. In 2022, Chat-GPT, a flagship of ANN use, worked with billions of adjustable parameters.

Curiously, this number has reversed its standing. In earlier times, such parameters indicated potential weaknesses and shortcomings. Hence, using a small number of adjustable parameters counted as sign of quality.

6. One can argue that speaking of learning is misleading because learning is among those actions that computers cannot do; or this is how the likes of Dreyfus (1972) put it in their controversial discussions over artificial intelligence.

7. We omit all the technical details because they do not matter here. Two informed and critical outlooks on AI that can be highly recommended are Smith (2019) and Russell (2019).

8. There is a controversial discussion in the community on whether and how the adjusted parameters induce representation in some way.

9. The obvious exception being the servers of the software giants.

10. Broad and controversial discussions around representation in philosophy of science actively engage in trying to pin down what it is that must be right.

11. Admittedly, experts with these tools usually develop standards of what counts as transparent that do not always match those developed by other persons.

12. The allegedly pure prediction culture is also a hybrid because it draws on data that are deeply entrenched into culture such as labeled pictures, translated texts, and so forth.

References

Aachener Beschlüsse. 1895. "Vorstandsrat des VDI, Bericht über Ingenieurslaboratorien (Stellungnahmen der Bezirke) und nachfolgend Aachener Beschlüsse." *Zeitschrift des Vereins Deutscher Ingenieure* 39: 1211–1216.

Achinstein, Peter. 1994. "Explanation vs. Prediction: Which Carries More Weight?" *PSA: Proceedings of the Biennial Meeting of the Philosophy of Science Association* 1994 (2): 156–164.

Adams, Douglas. 1979. *The Hitchhiker's Guide to the Galaxy*. London: Pan Books.

Agar, Jon. 2003. *The Government Machine. A Revolutionary History of the Computer*. Cambridge, MA: MIT Press.

Agrawal, Ajay, Joshua Gans, and Avi Goldfarb. 2018. *Prediction Machines. The Simple Economics of Artificial Intelligence*. Boston, MA: Harvard Business Review Press.

Akera, Atsushi. 2007. *Calculating a Natural World. Scientists, Engineers, and Computers during the Rise of U.S. Cold War Research*. Cambridge, MA: MIT Press.

Alder, Ken. 1997. *Engineering the Revolution. Arms and Enlightenment in France, 1763–1815*. Princeton, NJ: Princeton University Press.

Aldous, David, and James Allen Fill. 2002. *Reversible Markov Chains and Random Walks on Graphs*. https://www.stat.berkeley.edu/~aldous/RWG/book.pdf.

Allen, H. Julian, and A. J. Eggers, Jr. 1958. *A Study of the Motion and Aerodynamic Heating of Ballistic Missiles Entering the Earth's Atmosphere at High Supersonic Speeds*. No. 1381. NACA Report. Ames Aeronautical Laboratory.

Amsterdamski, Stefan. 1975. *Between Experience and Metaphysics: Philosophical Problems of the Evolution of Science*. Dordrecht, Netherlands: Reidel.

Anderson, J. D. 2009. "Motivation: An Example." In *Computational Fluid Dynamics*, edited by J. F. Wendt, 3rd ed., 3–14. Berlin and Heidelberg: Springer. https://doi.org/10.1007/978-3-540-85056-4.

Andersson, Jenny. 2012. "The Great Future Debate and the Struggle for the World." *The American Historical Review* 117 (5): 1411–1430.

Andersson, Jenny. 2018. *The Future of the World. Futurology, Futurists, and the Struggle for the Post-Cold War Imagination*. Oxford: Oxford University Press.

Ansoff, H. Igor, and Dennis P. Slevin. 1968. "An Appreciation of Industrial Dynamics." *Management Science* 14: 383–397.

Arend, Gerhard. 1998. *Die mechanik des Niccolò Tartaglia im kontext der zeitgenössischen erkenntnis- und wissenschaftstheorie*. München, Germany: Institut für Geschichte der Naturwissenschaften.

Arendt, Hannah. 1951. *The Origins of Totalitarianism*. New York: Schocken.

Arendt, Hannah. 1958. *The Human Condition*. Chicago: University of Chicago Press.

Ashford, Oliver M. 1985. *Prophet—or Professor? The Life and Work of Lewis Fry Richardson*. Bristol, UK: Adam Hilger.

Ashley, Richard K. 1983. "The Eye of Power: The Politics of World Modeling." *International Organization* 37 (3): 495–535.

Aspray, William, ed. 1990a. *Computing before Computers*. Ames: Iowa State University Press.

Aspray, William, ed. 1990b. *John von Neumann and the Origins of Modern Computing*. Cambridge, MA: MIT Press.

Aubin, David. 2017. "Ballistics, Fluid Mechanics, and Air Resistance at Gâvre, 1829–1915: Doctrine, Virtues, and the Scientific Method in a Military Context." *Archive for the History of Exact Sciences* 71: 509–542.

Bach, Carl. 1889. "Rezension Tetmajer." *Zeitschrift des Vereins Deutscher Ingenieure* 33: 452–455, 473–479.

Bach, Carl. 1891. "Ein üblicher Fehler bei gewissen hydraulischen Rechnungen." *Zeitschrift des Vereins Deutscher Ingenieure* 35: 474–476.

Bach, Carl. 1892. "Versuche über den Widerstand von Nietverbindungen gegen Gleiten." *Zeitschrift des Vereins Deutscher Ingenieure* 36: 1141–1148.

Bach, Carl. 1895. "Diskussionsbeitrag zu Aachener Beschlüssen im Protokoll der 36. VDI Hauptversammlung." *Zeitschrift des Vereins Deutscher Ingenieure* 39: 1215.

Bach, Carl. 1896. "Über Ingenieurserziehung. Rede bei der Sitzung des Württembergischen Bezirksvereins 2.1.1896." *Zeitschrift des Vereins Deutscher Ingenieure* 40: 268–269.

Bach, Carl. 1898. "Ermittlung der Zug-und Druckelastizität an dem gleichen Versuchskörper." *Zeitschrift des Vereins Deutscher Ingenieure* 42: 35–40.

Bach, Carl. 1899. *Die Maschinen–Elemente. Ihre Berechnung und Konstruktion mit Rücksicht auf die neueren Versuche*. 7th ed. 2 vols. Leipzig, Germany: Kröner.

Bach, Carl. 1920. *Elastizität und Festigkeit. Die für die Technik wichtigsten Sätze und deren erfahrungsmäßige Grundlage*. 8th ed. Berlin, Germany: Julius Springer.

Bach, Carl. 1926. *Mein Lebensweg und meine Tätigkeit. Eine Skizze*. Berlin, Germany: Julius Springer.

Bamford, Greg. 1993. "Popper's Explications of Ad Hocness: Circularity, Empirical Content, and Scientific Practice." *The British Journal for the Philosophy of Science* 44 (3): 335–355.

Barden, Christopher J., and Henry F. Schaefer, III. 2000. "Quantum Chemistry in the 21st Century." *Pure and Applied Chemistry* 72 (8): 1405–1423.

Barnett, Janet Heine. 2009. "Mathematics Goes Ballistic: Benjamin Robins, Leonhard Euler, and the Mathematical Education of Military Engineers." *British Society for the History of Mathematics Bulletin* 24: 92–104.

Bashforth, Francis. 1890. *The Bashforth Chronograph*. Cambridge: Cambridge University Press.

Baumann, Richard. 1917. "Wissenschaft, Geschäftsgeist und Hookesches Gesetz." *Zeitschrift des Vereins Deutscher Ingenieure* 61: 117–124.

Bedau, Mark A. 1997. "Weak Emergence." *Philosophical Perspectives* 11: 375–399.

Bedau, Mark A. 2002. "Downward Causation and the Autonomy of Weak Emergence." *Principia: An International Journal of Epistemology* 6 (1): 5–50.

Bélisle, Claude J. P., H. Edwin Romeijn, and Robert L. Smith. 1993. "Hit-and-Run Algorithms for Generating Multivariate Distributions." *Mathematics of Operations Research* 18: 255–266.

Bennett, Stuart. 1979. *A History of Control Engineering, 1800–1930*. London: Peter Peregrinus.

Berlin, Isaiah. 1953. *The Hedgehog and the Fox*. London: Weidenfeld and Nicolson.

Biagioli, Mario. 1989. "The Social Status of Italian Mathematicians, 1450–1600." *History of Science* 27: 41–94.

Biagioli, Mario. 1993. *Galileo, Courtier: The Practice of Science in the Culture of Absolutism*. Chicago: University of Chicago Press.

Bickelhaupt, F. Matthias, and Evert Jan Baerends. 2000. "Kohn-Sham Density Functional Theory: Predicting and Understanding Chemistry." In *Reviews in Computational Chemistry*. Vol. 15, edited by Kenny B. Lipkowitz and Donald B. Boyd, 1–86. Hoboken, NJ: Wiley.

Bjerknes, Vilhelm. 1904. "Das Problem Der Wettervorhersage, Betrachtet Vom Standpunkte Der Mechanik Und Der Physik." *Meteorologische Zeitschrift* 21: 1–7.

Bloomfield, Brian P. 1986. *Modelling the World: The Social Constructions of Systems Analysts*. Oxford: Basil Blackwell.

Böhme, Gernot, Wolfgang van den Daele, and Wolfgang Krohn. 1978. "The 'Scientification' of Technology." In *The Dynamics of Science and Technology*, edited by Wolfgang Krohn, Edwin T. Layton, Jr., and Peter Weingart, 219–250. Dordrecht, Netherlands: Reidel.

Bolcer, John D., and Robert B. Hermann. 1994. "The Development of Computational Chemistry in the United States." In *Reviews in Computational Chemistry*. Vol. 5, edited by Kenny B. Lipkowitz and Donald B. Boyd, 1–64. Hoboken, NJ: Wiley.

Borries, August von. 1895. "Ingenieurslaboratorien." *Zeitschrift des Vereins Deutscher Ingenieure* 39: 1212–1215.

Bos, Henk J. M. 1980. "Mathematics and Rational Mechanics." In *The Ferment of Knowledge. Studies in the Historiography of Eighteenth Century Science*, edited by George Sebastian Rousseau and Roy Porter, 327–355. Cambridge: Cambridge University Press.

Box, George E. P. 1983. "An Apology for Ecumenism in Statistics." In *Scientific Inference, Data Analysis, and Robustness*, edited by George E. P. Box, T. Leonard, and C. F. Wu, 51–84. New York: Academic Press.

Boys, Frank. 1950. "Electronic Wave Functions I. A General Method of Calculation for the Stationary States of Any Molecular System." *Proceedings of the Royal Society A* 200 (1063): 542–554.

Boys, Frank, and G. B. Cook. 1960. "Mathematical Problems in the Complete Quantum Prediction of Chemical Phenomena." *Reviews of Modern Physics* 32 (2): 285–294.

Bremer, Stuart A. 1977. *Simulated Worlds: A Computer Model of National Decision-Making*. Princeton, NJ: Princeton University Press.

Brown, Gary I. 1991. "The Evolution of the Term 'Mixed Mathematics.'" *Journal of the History of Ideas* 52: 81–102.

Brush, Stephen G. 1994. "Dynamics of Theory Change: The Role of Predictions." *PSA: Proceedings of the Biennial Meeting of the Philosophy of Science Association* 1994 (2): 133–145.

Bugos, Glenn E. 2014. *Atmosphere of Freedom. 75 Years at NASA Ames Research Center*. NASA SP-2014-4314. Washington, DC: NASA History Office. https://history.arc.nasa.gov/hist_pdfs/bugos_nasa_sp2014_4314.pdf.

Burr, William H. 1893. "The Ideal Engineering Education." *Scientific American Supplement* 920–921: 14699–14701, 14715–14716.

Calvert, Monte A. 1967. *The Mechanical Engineer in America, 1830–1910. Professional Cultures in Conflict*. Baltimore, MD: Johns Hopkins University Press.

Carlin, Bradley P. 2004. "Whither Applied Bayesian Inference?" In *Applied Bayesian Modeling and Causal Inference from Incomplete-Data Perspectives*, edited by Andrew Gelman and Xiao-Li Meng, 279–284. New York: John Wiley and Sons.

Carlin, Bradley P., and Thomas A. Louis. 2000. *Bayes and Empirical Bayes Methods for Data Analysis*. New York: Chapman and Hall/CRC Press.

Cartwright, Nancy. 2020. "Why Trust Science? Reliability, Particularity and the Tangle of Science." *Proceedings of the Aristotelian Society* 120 (3): 237–252.

Ceruzzi, Paul. 2003. *A History of Modern Computing*. Cambridge, MA: MIT Press.

Chalmers, Alan. 2013. *What Is This Thing Called Science?* 4th ed. New York: Open University Press.

Chang, Hasok. 2004. *Inventing Temperature*. New York: Oxford University Press.

Chen, Ming-Hui, and Bruce W. Schmeiser. 1996. "General Hit-and-Run Monte Carlo Sampling for Evaluating Multidimensional Integrals." *Operations Research Letters* 19: 161–169.

Clark, Christopher. 2019. *Time and Power: Visions of History in German Politics, from the Thirty Years' War to the Third Reich*. Princeton, NJ: Princeton University Press.

Clark, Wesley. 1988. "The LINC Was Early and Small." In *A History of Personal Workstations*, edited by Adele Goldberg, 345–400. New York: ACM Press.

Clarke, Arthur C. 1973. *Profiles of the Future*. New York: Harper & Row.

Clifford, Peter. 1993. "Discussion Statement, in: Discussion on the Meeting on the Gibbs Sampler and Other Markov Chain Monte Carlo Methods." *Journal of the Royal Statistical Society B* 55 (1): 53–102.

Cole, H. S. D., and R. C. Curnow. 1973. "An Evaluation of the World Models." In *Models of Doom—A Critique of the Limits to Growth*, edited by H. S. D. Cole, Christopher Freeman, Marie Jahoda, and K. L. R. Pavitt, 108–134. London: Chatto and Windus.

Cole, H. S. D., Christopher Freeman, Marie Jahoda, and K. L. R. Pavitt, eds. 1973. *Models of Doom—A Critique of the Limits to Growth*. London: Chatto and Windus.

Corfield, David, and Jon Williamson, eds. 2001. *Foundations of Bayesianism*. Dordrecht, Netherlands: Kluwer Academic Publishers.

Coulson, Charles A. 1947. "The Meaning of Resonance in Quantum Chemistry." *Endeavour* 6: 42–47.

Coulson, Charles A. 1960. "Present State of Molecular Structure Calculations." *Reviews of Modern Physics* 32: 170–177.

Coulson, Charles A. 1970. "Recent Developments in Valence Theory." *Pure and Applied Chemistry* 24: 257–287.

Crowther-Heyck, Hunter, and Herbert A. Simon. 2005. *The Bounds of Reason in Modern America*. Baltimore, MD: Johns Hopkins University Press.

Cuomo, Serafina. 1997. "Shooting by the Book: Notes on Niccolò Tartaglia's Nova Scientia." *History of Science* 35: 155–188.

Cuomo, Serafina. 1998. "Niccolò Tartaglia, Mathematics, Ballistics and the Power of Possession of Knowledge." *Endeavour* 22 (1): 31–35.

Daston, Lorraine. 1988. *Classical Probability in the Enlightenment*. Princeton, NJ: Princeton University Press.

Daston, Lorraine, and Peter Galison. 2007. *Objectivity*. New York: Zone Books.

Datta, Gauri S., and Trevor J. Sweeting. 2004. "Probability Matching Priors." In *Bayesian Thinking: Modeling and Computation*, edited by Dipak Dey and C. R. Rao, Handbook of Statistics. Vol. 25, 91–114. Amsterdam, Netherlands: Elsevier.

Day, Richard H. 1974. "On System Dynamics." *Behavioral Science* 19: 260–271.

De Chadarevian, Soraya, and Nick Hopwood. 2004. *Models. The Third Dimension of Science*. Stanford, CA: Stanford University Press.

Dear, Peter. 1995. *Discipline and Experience. The Mathematical Way in the Scientific Revolution*. Chicago: University of Chicago Press.

Dear, Peter. 2006. *The Intelligibility of Nature. How Science Makes Sense of the World*. Chicago: University of Chicago Press.

Denny, Mark. 2011. *Their Arrows Will Darken the Sun*. Baltimore, MD: Johns Hopkins University Press.

Desrosières, Alain. 2000. *La politique des grands nombres: Histoire de la raison statistique*. Paris: La découverte.

Deutsch, Karl W., Bruno Fritsch, Helio Jaguaribe, and Andrei S. Markovits, eds. 1977. *Problems of World Modeling. Political and Social Implications*. Cambridge, MA: Ballinger.

Diaconis, Persi. 2009. "The Markov Chain Monte Carlo Revolution." *Bulletin of the American Mathematical Society* 46 (2): 179–205.

Diaconis, Persi. 2013. "Some Things We've Learned (about Markov Chain Monte Carlo)." *Bernoulli* 19 (4): 1294–1305.

Diaconis, Persi, and David Freedman. 1986. "On the Consistency of Bayes Estimates (with Discussion)." *Annals of Statistics* 14: 1–67.

Dienel, Hans-Liudger. 1993. "Der Münchner Weg im Theorie-Praxis-Streit um die Emanzipation des wissenschaftlichen Maschinenbaus." In *Technische Universität München. Annäherungen an ihre Geschichte*, edited by Ulrich Wengenroth, 87–115. München, Germany: Technische Universität München.

Dirac, Paul A. M. 1929. "Quantum Mechanics of Many Electron Systems." *Proceedings of the Royal Society A* 123: 713–733.

Ditchen, Henryk. 2016. "Geschichte der Materialforschung in Europa." In *Geschichte und Praxis der Materialforschung*, edited by Klaus Hentschel and Josef Webel, 53–101. Diepholz, Germany: NGT-Verlag.

Drake, Stillman, and I. E. Drabkin. 1969. *Mechanics in Sixteenth-Century Italy. Selections from Tartaglia, Benedetti, Guido Ubaldo, and Galileo.* Madison: University of Wisconsin Press.

Dreyfus, Hubert. 1972. *What Computers Can't Do. The Limits of Artificial Intelligence.* New York: Harper & Row.

Durand, William F. 1939. "Dr. Robert Henry Thurston's Eighteen Years at Cornell." *Science* 90 (2346): 547–552.

Dyck, Walther von. 1898. "Zur Frage der Ingenieurausbildung." *Zeitschrift des Vereins Deutscher Ingenieure* 42: 1276–1278.

Earman, John. 1992. *Bayes or Bust? A Critical Examination of Bayesian Confirmation Theory*. Cambridge, MA: MIT Press.

Edwards, Paul N. 1996. *The Closed World. Computers and the Politics of Discourse in Cold War America*. Cambridge, MA: MIT Press.

Edwards, Paul N. 2000. "The World in a Machine: Origins and Impacts of Early Computerized Global Systems Models." In *Systems, Experts, and Computers: The Systems Approach in Management and Engineering, World War II and After*, edited by Thomas P. Hughes and Agatha C. Hughes, 221–254. Cambridge, MA: MIT Press.

Edwards, Paul N. 2010. *A Vast Machine. Computer Models, Climate Data, and the Politics of Global Warming*. Cambridge, MA: MIT Press.

Efron, Bradley. 2005. "Bayesians, Frequentists, and Scientists." *Journal of the American Statistical Association* 100 (469): 1–5.

Efron, Bradley, and Trevor J. Hastie. 2016. *Computer Age Statistical Inference: Algorithms, Evidence, and Data Science*. Cambridge: Cambridge University Press.

Ekholm, Karin J. 2010. "Tartaglia's Ragioni: A Maestro d'abaco's Mixed Approach to the Bombardier's Problem." *British Journal for the History of Science* 43 (2): 181–207.

Engel, Eberhard, and Reiner M. Dreizler. 2011. *Density Functional Theory. An Advanced Course*. Berlin and Heidelberg: Springer.

Erlichson, Herman. 1998. "How Galileo Solved the Problem of Maximum Projectile Range without the Calculus." *European Journal of Physics* 19: 251–257.

Ernst, Adolf. 1894. "Maschinenbaulaboratorien." *Zeitschrift des Vereins Deutscher Ingenieure* 38: 1351–1362.

Euler, Leonhard. 1922a. "Neue Grundsätze der Artillerie aus dem Englischen des Herrn Benjamin Robins übersetzt und mit vielen Anmerkungen." In *Opera Omnia, Second Series XIV*, 1–409. Berlin: B. G. Teubner-Verlag.

Euler, Leonhard. 1922b. "Recherches sur la veritable courbe que décrivent les corps jettés dans l'air uu dans un autre fluide quelconque." In *Opera Omnia, Second Series XIV*, 413–47. Berlin: B. G. Teubner-Verlag.

Exner, Felix Maria von. 1923. "Book Review." *Meteorologische Zeitschrift* 40: 189–191.

Feyerabend, Paul. 1999. *The Conquest of Abundance*. Chicago: University of Chicago Press.

Fischer, Charlotte Froese. 2003. *Douglas Rayner Hartree. His Life in Science and Computing*. Singapore: World Scientific Publisher.

Fleck, Ludwik. 1979. *Genesis and Development of a Scientific Fact*. Chicago: University of Chicago Press.

Föppl, August. 1897. "Ziele und methoden der technischen mechanik." *Jahresbericht der Deutschen mathematiker–Vereinigung* 6: 99–110.

Foresman, James B., and Aeleen Frisch. 1993. *Exploring Chemistry with Electronic Structure Methods. A Guide to Using Gaussian*. Pittsburgh, PA: Gaussian.

Forrester, Jay W. 1961. *Industrial Dynamics*. Cambridge, MA: MIT Press.

Forrester, Jay W. 1969. *Urban Dynamics*. Cambridge, MA: MIT Press.

Forrester, Jay W. 1971. *World Dynamics*. Cambridge, MA: Wright-Allen Press.

Forrester, Jay W. 1980. *Principles of Systems*. 9th ed. Cambridge: MIT Press.

Forsythe, George E. 1959. "The Role of Numerical Analysis in an Undergraduate Program." *American Mathematical Monthly* 66: 651–662.

Freeman, Christopher. 1973. "Malthus with a Computer." In *Models of Doom—A Critique of the Limits to Growth*, edited by H. S. D. Cole, Christopher Freeman, Marie Jahoda, and K. L. R. Pavitt, 5–13. London: Chatto and Windus.

Frenkel, Daan. 2013. "Simulations: The Dark Side." *The European Physical Journal Plus* 128 (10): 1–21. https://doi.org/10.1140/epjp/i2013-13010-8.

Friedman, Robert M. 1989. *Appropriating the Weather—Vilhelm Bjerknes and the Construction of Modern Meteorology*. New York: Cornell University Press.

Friedman, Walter A. 2014. *Fortune Tellers. The Story of America's First Economic Forecasters*. Princeton, NJ: Princeton University Press.

Friedrich, Simon, Robert V. Harlander, and Koray Karaca. 2014. "Philosophical Perspectives on *Ad Hoc* Hypotheses and the Higgs Mechanism." *Synthese* 191: 3897–3917.

Galileo. 1974. *Two New Sciences*. Madison: University of Wisconsin Press.

Galison, Peter. 1997. *Image and Logic: A Material Culture of Microphysics*. Chicago: University of Chicago Press.

Gavroglu, Kostas, and Ana I. Simões. 1994. "The Americans, the Germans, and the Beginnings of Quantum Chemistry: The Confluence of Diverging Traditions." *Historical Studies in the Physical and Biological Sciences* 25 (1): 47–110.

Gavroglu, Kostas, and Ana I. Simões. 2012. *Neither Physics nor Chemistry: A History of Quantum Chemistry*. Boston, MA: MIT Press.

Gelfand, Alan E., and Adrian F. M. Smith. 1990. "Sampling-Based Approaches to Calculating Marginal Densities." *Journal of the American Statistical Association* 85: 398–409.

Gelman, Andrew. 2011. "Bayesian Statistical Pragmatism." *Statistical Science* 26 (1): 10–11.

Geman, Stuart, and Donald Geman. 1984. "Stochastic Relaxation, Gibbs Distributions, and the Bayesian Restoration of Images." *IEEE Transactions on Pattern Analysis and Machine Intelligence* 6: 721–741.

Geymonat, Ludovico. 1965. *Galileo Galilei: A Biography and Inquiry into His Philosophy of Science*. New York: McGraw-Hill.

Giles, Jim. 2004. "Software Company Bans Competitive Users." *Nature* 429: 231. https://doi.org/10.1038/429231a.

Gilks, Walter R., A. Thomas, and David J. Spiegelhalter. 1994. "A Language and Program for Complex Bayesian Modelling." *The Statistician* 43 (1): 169–177.

Gill, Jeff. 2008. *Bayesian Methods. A Social and Behavioral Sciences Approach*. 2nd ed. Cambridge: Cambridge University Press.

Gispen, Kees. 1989. *New Profession, Old Order. Engineers and German Society, 1815–1914*. Cambridge: Cambridge University Press.

Gitelman, Lisa, ed. 2013. *"Raw Data" Is an Oxymoron*. Cambridge, MA: MIT Press.

Gluchoff, Alan. 2011. "Artillerymen and Mathematicians: Forest Ray Moulton and Changes in American Exterior Ballistics, 1885–1934." *Fuel and Energy Abstracts* 38: 506–547.

Good, Irving John. 1983. *Good Thinking*. Minneapolis: University of Minnesota Press.

Goodman, Steven N. 2011. "Discussion of 'Statistical Inference: The Big Picture' by R. E. Kass." *Statistical Science* 26 (1): 12–14. https://doi.org/10.1214/11-STS337A.

Gosman, A. D., W. M. Pun, Akshai K. Runchal, D. Brian Spalding, and M. Wolfshtein. 1969. *Heat and Mass Transfer in Recirculating Flows*. London: Academic Press.

Grashof, Franz. 1875. *Theoretische Maschinenlehre*. Vol. 3. Leipzig and Berlin: Voss.

Grattan-Guinness, Ivor. 1993. "The Ingénieur Savant, 1800–1830. A Neglected Figure in the History of French Mathematics and Science." *Science in Context* 6 (2): 405–433.

Greenland, Sander. 2010. "Comment: The Need for Syncretism in Applied Statistics." *Statistical Science* 25 (2): 158–161.

Grier, David Alan. 2001. "Dr. Veblen Takes a Uniform: Mathematics in the First World War." *American Mathematical Monthly* 108: 922–931.

Grier, David Alan. 2005. *When Computers Were Human*. Princeton, NJ: Princeton University Press.

Grünbaum, Adolf. 1976. "*Ad Hoc* Auxiliary Hypotheses and Falsificationism." *British Journal for the Philosophy of Science* 27: 329–362.

Guetzkow, Harold, and Joseph J. Valadez, eds. 1981. *Simulated International Processes: Theories and Research in Global Modeling*. Beverly Hills, CA: Sage.

Hacking, Ian. 1982. "Language, Truth and Reason." In *Rationality and Relativism*, edited by M. Hollis and S. Lukes, 48–66. Oxford: Blackwell.

Hacking, Ian. 1990. *The Taming of Chance*. Cambridge: Cambridge University Press.

Hacking, Ian. 2001. *An Introduction to Probability and Inductive Logic*. Cambridge: Cambridge University Press.

Häggström, Olle. 2002. *Finite Markov Chains and Algorithmic Applications*. Cambridge: Cambridge University Press.

Haigh, Thomas, Mark Priestley, and Crispin Rope. 2016. *ENIAC in Action. Making and Remaking the Modern Computer*. Cambridge, MA: MIT Press.

Hall, A. Rupert. 2009. *Ballistics in the Seventeenth Century*. Cambridge: Cambridge University Press.

Hall, Bert S. 1997. *Weapons and Warfare in Renaissance Europe: Gunpowder, Technology, and Tactics*. Baltimore, MD: Johns Hopkins University Press.

Halpern, Paul. 2000. *The Pursuit of Destiny: A History of Prediction*. Cambridge, MA: Perseus.

Handy, Nicolas C., John A. Pople, and Isaiah Shavitt. 1996. "Samuel Francis Boys." *Journal of Physical Chemistry* 100: 6007–6016.

Hardin, Garrett. 1972. "The Tragedy of the Commons." *Science* 162 (3859): 1243–1248.

Harlow, Francis H. 2004. "Fluid Dynamics in Group T-3, Los Alamos National Laboratory: (LA-UR-03–3852)." *Journal of Computational Physics* 195 (2): 414–433.

Harlow, Francis H., John P. Shannon, and J. Eddie Walsh. 1965. "Liquid Waves by Computer." *Science* 149 (3688): 1092–1093.

Harper, Kristin C. 2008. *Weather by the Numbers: The Genesis of Modern Meteorology*. Cambridge, MA: MIT Press.

Harris, Martha L. 2008. "Chemical Reductionism Revisited: Lewis, Pauling, and the Physico-Chemical Nature of the Chemical Bond." *Studies in History and Philosophy of Science* 39: 78–90.

Hartmann, Heinrich, and Jakob Vogel. 2010. *Zukunftswissen. Prognosen in Wirtschaft, Politik und Gesellschaft seit 1900*. Frankfurt, Germany: Campus Verlag.

Hartog, François. 2015. *Regimes of Historicity: Presentism and Experiences of Time*. New York: Columbia University Press.

Hartree, Douglas R. 1949. *Calculating Instruments and Machines*. Urbana: University of Illinois Press.

Hartree, Douglas R. 1958. *Numerical Analysis*. 2nd ed. Oxford: Oxford University Press.

Hartree, Douglas R. 1984. *Calculating Machines. Recent and Prospective Developments and Their Impact on Mathematical Physics*. Cambridge, MA: MIT Press.

Hashagen, Ulf. 2003. *Walther von Dyck (1856–1934). Mathematik, Technik und Wissenschaftsorganisation an der TH München*. Stuttgart, Germany: Franz Steiner Verlag.

Hastings, W. K. 1970. "Monte Carlo Sampling Methods Using Markov Chains and Their Applications." *Biometrika* 57 (1): 97–109.

Heaviside, Oliver. 1892. "On Operators in Physical Mathematics. Part I." *Proceedings of the Royal Society* 52: 504–529.

Heaviside, Oliver. 1893. "On Operators in Physical Mathematics. Part II." *Proceedings of the Royal Society* 54: 105–143.

Heitler, Walter, and Fritz London. 1927. "Wechselwirkung neutraler Atome und homoeopolare Bindung nach der Quantenmechanik." *Zeitschrift Für Physik* 44: 455–472.

Hensel, Susann. 1989. "Zur mathematischen Ausbildung der Ingenieure an den Technischen Hochschulen Deutschlands im letzten Drittel des 19. Jahrhunderts." *NTM Journal of the History of Science, Technology and Medicine* 26: 33–47.

Hensel, Susann. 1991. "Der Technikwissenschaftler Carl Julius Bach (1847–1931) und die Frage der Mathematischen Ausbildung der Ingenieure." *NTM Journal of the History of Science, Technology and Medicine* 28: 231–257.

Hensel, Susann, Karl-Norbert Ihmig, and Michael Otte. 1989. *Mathematik und Technik im 19. Jahrhundert: Soziale Auseinandersetzung und philosophische Problematik.* Goettingen, Germany: Vandenhoeck & Ruprecht.

Hepler-Smith, Evan. 2018. "'A Way of Thinking Backwards': Computing and Method in Synthetic Organic Chemistry." *Historical Studies in the Natural Sciences* 48 (3): 300–337.

Herrera, Amílcar Oscar, Hugo D. Scolnik, Graciela Chichilnisky, Gilberto C. Gallopin, and Jorge Enrique Hardoy. 1976. *Catastrophe or New Society? A Latin American World Model.* Ottawa, ON, Canada: International Development Research Center.

Heyck, Hunter. 2015. *The Age of System. Understanding the Development of Modern Social Science.* Baltimore, MD: Johns Hopkins University Press.

Heymann, Matthias, Gabriele Gramelsberger, and Martin Mahony, eds. 2017. *Cultures of Prediction in Atmospheric and Climate Science.* London: Routledge.

Hocquet, Alexandre, and Frédéric Wieber. 2021. "Epistemic Issues in Computational Reproducibility: Software as the Elephant in the Room." *European Journal for Philosophy of Science* 11 (2): 1–20. https://doi.org/10.1007/s13194-021-00362-9.

Hohenberg, Pierre, and Walter Kohn. 1964. "Inhomogeneous Electron Gas." *Physical Review* 136 (3B): B864–B871. https://doi.org/10.1103/PhysRev.136.B864.

Hounshell, David A. 1980. "Edison and the Pure Science Ideal in Nineteenth-Century America." *Science* 207 (4431): 612–617.

Howson, Colin, and Peter Urbach. 1993. *Scientific Reasoning: The Bayesian Approach.* 2nd ed. Chicago: Open Court.

Hughes, Agatha C., and Thomas P. Hughes. 2000. *Systems, Experts, and Computers: The Systems Approach in Management and Engineering, World War II and After.* Cambridge, MA: MIT Press.

Hughes, Barry B. 1980. *World Modeling: The Mesarovic-Pestel World Model in the Context of Its Contemporaries.* Lexington, MA: Lexington Books.

Hughes, R. I. G. 1997. "Models and Representation." *Philosophy of Science* 64 (Proceedings): S325–S336.

Hughes, R. I. G. 1999. "The Ising Model, Computer Simulation, and Universal Physics." In *Models as Mediators*, edited by Mary S. Morgan and Margaret Morrison. Cambridge: Cambridge University Press.

Hughes, Thomas P. 1989. *American Genesis. A Century of Invention and Technological Enthusiasm 1870–1970.* New York: Viking Penguin.

Hughes, Thomas P. 1998. *Rescuing Prometheus: Four Monumental Projects That Changed the Modern World.* New York: Pantheon.

Hughes, Thomas P. 2004. *Human-Built World. How to Think about Technology and Culture*. Chicago: University of Chicago Press.

Humphreys, Paul. 2004. *Extending Ourselves. Computational Science, Empiricism, and Scientific Method*. New York: Oxford University Press.

Humphreys, Paul. 2009. "The Philosophical Novelty of Computer Simulations Methods." *Synthese* 169: 615–626.

Hunt, Bruce J. 1991. *The Maxwellians*. Ithaca, NY: Cornell University Press.

Hunt, J. C. R. 1998. "Lewis Fry Richardson and His Contributions to Mathematics, Meteorology, and Models of Conflict." *Annual Review of Fluid Mechanics* 30 (1): xiii–xxxvi.

Hyman, Anthony. 1985. *Charles Babbage, Pioneer of the Computer*. Princeton, NJ: Princeton University Press.

Ihmig, Karl-Norbert. 1989. "Das Verhältnis von Mathematik und Kinematik bei Franz Reuleaux." In *Mathematik und Technik im 19. Jahrhundert: Soziale Auseinandersetzung und philosophische Problematik*, edited by Susann Hensel, Karl-Norbert Ihmig, and Michael Otte, 112–148. Goettingen, Germany: Vandenhoeck & Ruprecht.

Jantsch, Erich. 1967. *Technological Forecasting in Perspective*. Paris: OECD Publications.

Johnson, Ann. 2004. "From Boeing to Berkeley: Civil Engineers, the Cold War and the Development of Finite Element Analysis." In *Growing Explanations: Historical Perspectives on the Sciences of Complexity*, edited by M. Norton Wise, 133–158. Durham, UK: Duke University Press.

Johnson, Ann. 2017. "Rational and Empirical Cultures of Prediction." In *Mathematics as a Tool*, edited by Johannes Lenhard and Martin Carrier, 23–35. Boston Studies in the History and Philosophy of Science. Vol. 327. Cham, Switzerland: Springer.

Johnson, Ann, and Johannes Lenhard. 2011. "Toward a New Culture of Prediction: Computational Modeling in the Era of Desktop Computing." In *Science Transformed? Debating Claims of an Epochal Break*, edited by Alfred Nordmann, Hans Radder, and Gregor Schiemann, 189–199. Pittsburgh: University of Pittsburgh Press.

Johnson, William. 1992. "Encounters between Robins, and Euler and the Bernoullis; Artillery and Related Subjects." *International Journal of Mechanical Sciences* 34 (8): 651–679.

Kaiser, David. 2005. *Drawing Theories Apart. The Dispersion of Feynman Diagrams in Postwar Physics*. Chicago: University of Chicago Press.

Kass, Robert E. 2011. "Statistical Inference: The Big Picture." *Statistical Science* 26 (1): 1–9.

Kass, Robert E., Bradley P. Carlin, Andrew Gelman, and Radford M. Neal. 1998. "Markov Chain Monte Carlo in Practice: A Roundtable Discussion." *The American Statistician* 52 (2): 93–100.

Kieseppä, I. A. 1997. "Akaike Information Criterion, Curve-Fitting, and the Philosophical Problem of Simplicity." *British Journal for the Philosophy of Science* 48 (1): 21–48.

Klein, Felix. 1894. *The Evanston Colloquium.* New York: McMillan.

Klein, Felix. 1895. "Über die Beziehungen der neueren Mathematik zu den Anwendungen. Antrittsrede, gehalten am 25.10.1880 in Leipzig." *Zeitschrift für den naturwissenschaftlich-mathematischen Unterricht* 26: 535–540.

Klein, Felix. 1896a. "Die Anforderungen der Ingenieure und die Ausbildung der mathematischen Lehramtskandidaten." *Zeitschrift des Vereins Deutscher Ingenieure* 40: 980–987.

Klein, Felix. 1896b. "Über den Plan eines Physikalisch-Technischen Instituts an der Universität Göttingen. Sitzung Hannoverscher Bezirksverein des VDI am 6.12.1895." *Zeitschrift des Vereins Deutscher Ingenieure* 40: 102–107.

Klein, Felix. 1896c. "Über den Plan eines Physikalisch-Technischen Instituts an der Universität Göttingen. Sitzung Württembergischer Bezirksverein 3.7.1895, Berichterstattung über F. Kleins Denkschrift: Über die Gründung eines Physikalisch-Technischen Universitätsinstituts in Göttingen, nebst Stellungnahme basierend auf C. Bach." *Zeitschrift des Vereins Deutscher Ingenieure* 40: 75–77.

Klein, Felix. 1898. "Universität und Technische Hochschule." *Zeitschrift des Vereins Deutscher Ingenieure* 42: 1091–1094.

Klein, Felix. 1900a. "Allgemeines über angewandte Mathematik." In *Über angewandte Mathematik und Physik in ihrer Bedeutung für den Unterricht an den höheren Schulen,* edited by Felix Klein and Eduard Riecke, 15–25. Leipzig and Berlin: B. G. Teubner.

Klein, Felix. 1900b. "Über technische Mechanik." In *Über angewandte Mathematik und Physik in ihrer Bedeutung für den Unterricht an den höheren Schulen,* edited by Felix Klein and Eduard Riecke, 26–41. Leipzig and Berlin: B. G. Teubner.

Klein, Felix. 2016. *Elementary Mathematics from a Higher Standpoint.* Translated by Marta Menghini. Berlin and Heidelberg: Springer.

Klein, Ursula. 1999. "Techniques of Modelling and Paper Tools in Classical Chemistry." In *Models as Mediators: Perspectives on Natural and Social Science,* edited by Mary S. Morgan and Margaret Morrison, 146–167. New York: Cambridge University Press.

Kline, Ron A. 2018. "Mathematical Models of Technological and Social Complexity." In *Technology and Mathematics,* edited by Sven Ove Hanson, 285–303. Cham, Switzerland: Springer.

Kline, Ronald R. 1987. "Science and Engineering Theory in the Invention and Development of the Induction Motor, 1880–1900." *Technology and Culture* 28 (2): 283–313.

Kline, Ronald R. 1992. *Steinmetz. Engineer and Socialist.* Baltimore, MD: Johns Hopkins University Press.

Kline, Ronald R. 1995. "Construing 'Technology' as 'Applied Science': Public Rhetoric of Scientists and Engineers in the United States, 1880–1945." *Isis* 86 (2): 194–221.

Kline, Ronald R. 2015. *The Cybernetics Moment. Or Why We Call Our Age the Information Age.* Baltimore, MD: Johns Hopkins University Press.

Kohn, Walter, and Lu Jeu Sham. 1965. "Self-Consistent Equations Including Exchange and Correlation Effects." *Physical Review* 140 (4A): A1133–A1138.

König, Wolfgang. 1993. "Technical Education and Industrial Performance in Germany: A Triumph of Heterogeneity." In *Education, Technology and Industrial Performance in Europe, 1850–1939*, edited by Robert Fox and Anna Guagnini, 65–87. Cambridge: Cambridge University Press.

König, Wolfgang. 2014. *Der Gelehrte und der Manager. Franz Reuleaux (1829–1905) und Alois Riedler (1850–1936) in Technik, Wissenschaft und Gesellschaft.* Stuttgart, Germany: Franz Steiner Verlag.

Koselleck, Reinhart. 1981. "Modernity and the Planes of Historicity." *Economy and Society* 10 (2): 166–183.

Koselleck, Reinhart. 2004. *Futures Past. On the Semantics of Historical Time.* New York: Columbia University Press.

Koyré, Alexandre. 1957. *From the Closed World to the Infinite Universe.* Baltimore, MD: Johns Hopkins University Press.

Krishnan, Raghavachari, Michael J. Frisch, and John A. Pople. 1980. "Contribution of Triple Substitutions to the Electron Correlation Energy in Fourth Order Perturbation Theory." *Journal of Chemical Physics* 72 (1): 4244–4255.

Krüger, Lorenz, Lorraine Daston, and Michael Heidelberger, eds. 1987. *The Probabilistic Revolution.* 2 vols. Cambridge, MA: MIT Press.

Latour, Bruno. 1987. *Science in Action. How to Follow Scientists and Engineers through Society.* Cambridge, MA: Harvard University Press.

Latour, Bruno. 1990. "Drawing Things Together." In *Representation in Scientific Practice*, edited by Michael Lynch and Steve Woolgar, 19–68. Boston, MA: MIT Press.

Layton, Edwin T. Jr. 1971. "Mirror-Image Twins: The Communities of Science and Technology in Nineteenth-Century America." *Technology and Culture* 12 (4): 562–580.

Layton, Edwin T. Jr. 1986. *The Revolt of the Engineers. Social Responsibility and the American Engineering Profession.* Baltimore, MD: Johns Hopkins University Press.

Lejaeghere, Kurt, Gustav Bihlmayer, Torbjörn Björkman, Peter Blaha, Stefan Blügel, Volker Blum, Damien Caliste, et al. 2016. "Reproducibility in Density Functional Theory Calculations of Solids." *Science* 351 (6280): aad3000. https://doi.org/10.1126/science.aad3000.

Lenhard, Johannes. 2006. "Models and Statistical Inference: The Controversy between Fisher and Neyman–Pearson." *British Journal for the Philosophy of Science* 57: 69–91.

Lenhard, Johannes. 2007. "Computer Simulation: The Cooperation between Experimenting and Modeling." *Philosophy of Science* 74: 176–194.

Lenhard, Johannes. 2014. "Disciplines, Models, and Computers: The Path to Computational Quantum Chemistry." *Studies in History and Philosophy of Science Part A* 48: 89–96.

Lenhard, Johannes. 2019. *Calculated Surprises. A Philosophy of Computer Simulation.* New York: Oxford University Press.

Lenhard, Johannes. 2022. "A Transformation of Bayesian Statistics: Computation, Prediction, and Rationality." *Studies in History and Philosophy of Science* 92: 144–152.

Lenhard, Johannes, and Martin Carrier, eds. 2017. *Mathematics as a Tool. Tracing New Roles for Mathematics in the Sciences*. Boston Studies in the History and Philosophy of Science 327. Cham, Switzerland: Springer.

Lennox, James G. 1986. "Aristotle, Galileo, and the 'Mixed Sciences.'" In *Reinterpreting Galileo*, edited by William Wallace, 29–51. Washington, DC: Catholic University of America Press.

Leontief, Wassily W. 1977. *The Future of the World Economy: A United Nations Study.* New York: Oxford University Press.

Leplin, Jarrett. 1975. "The Concept of an Ad Hoc Hypothesis." *Studies in History and Philosophy of Science* 5: 309–345.

Levin, David A., Yuval Peres, and Elizabeth L. Wilmer. 2009. *Markov Chains and Mixing Times*. Providence, RI: American Mathematical Society.

Levin, Miriam R., ed. 2004. *Cultures of Control*. London: Routledge.

Licklider, Joseph C. R., George Shapiro, and Milton Rodgers. 1967. "Interactive Dynamic Modeling." In *Prospects for Simulation and Simulators of Dynamic Modeling*, 281–289. New York: Spartan Books.

Liepmann, Hans Wolfgang, and Anatol Roshko. 1957. *Elements of Gasdynamics*. New York: John Wiley and Sons.

Lilienfeld, Robert. 1978. *The Rise of Systems Theory. An Ideological Analysis*. New York: John Wiley and Sons.

Lindley, David V. 1965. *Introduction to Probability and Statistics from a Bayesian Viewpoint*. Cambridge: Cambridge University Press.

Lindsay, Robert K., Edward A. Feigenbaum, and Joshua Lederberg. 1980. *Applications of Artificial Intelligence to Chemical Inference. The Dendral Project*. New York: McGraw-Hill.

Lipkowitz, Kenny B., and Donald B. Boyd, eds. 1994. *Reviews in Computational Chemistry*. Vol. 5. New York: VCH Publishers.

Lipkowitz, Kenny B., and Donald B. Boyd, eds. 1997. *Reviews in Computational Chemistry*. Vol. 10. New York: VCH Publishers.

Lipkowitz, Kenny B., and Donald B. Boyd. 2000. "A Tribute to the Halcyon Days of QCPE." In *Reviews in Computational Chemistry*. Vol. 15, edited by Kenny B. Lipkowitz and Donald B. Boyd, v–xvi. New York: VCH Publishers.

Long, Lyle N., and Howard Weiss. 1999. "The Velocity Dependence of Aerodynamic Drag: A Primer for Mathematicians." *American Mathematical Monthly* 106: 127–135.

Lövdin, Per-Olov. 1967. "Program." *International Journal of Quantum Chemistry* 1: 1–6.

Lucier, Paul. 2012. "The Origins of Pure and Applied Science in Gilded Age America." *Isis* 103 (3): 527–536.

Lundgreen, Peter. 1990. "Engineering Education in Europe and the U.S.A., 1750–1930: The Rise to Dominance of School Culture and the Engineering Professions." *Annals of Science* 47 (1): 33–75.

Lunn, David J., David J. Spiegelhalter, Andrew Thomas, and Nicky Best. 2009. "The BUGS Project: Evolution, Critique and Future Directions." *Statistics in Medicine* 28: 3049–3082.

Lunn, David J., Andrew Thomas, Nicky Best, and David J. Spiegelhalter. 2000. "WinBUGS—A Bayesian Modelling Framework: Concepts, Structure, and Extensibility." *Statistics and Computing* 10: 325–337.

Lynch, Peter. 1993. "Richardson's Forecast Factory: The 64 000 $ Question." *Meteorological Magazine* 122: 69–70.

Machamer, Peter. 1978. "Galileo and the Causes." In *New Perspectives on Galileo*, edited by Robert Butts and Joseph Pitt, 161–180. Dordrecht, Netherlands: Reidel.

Machamer, Peter. 1998. "Galileo's Machines, His Mathematics, and His Experiments." In *The Cambridge Companion to Galileo*, edited by Peter Machamer, 53–79. Cambridge: Cambridge University Press.

MacKenzie, Donald. 2000. "A Worm in the Bud? Computers, Systems, and the Safety-Case Problem." In *Systems, Experts, and Computers: The Systems Approach in Management and Engineering, World War II and After*, edited by Agatha C. Hughes and Thomas P. Hughes, 161–190. Cambridge, MA: MIT Press.

MacKenzie, Donald. 2001. *Mechanizing Proof. Computing, Risk, and Trust*. Cambridge, MA: MIT Press.

MacKenzie, Donald. 2006. *Engine, Not Camera. How Financial Models Shape Markets*. Cambridge, MA: MIT Press.

Maddox, John. 1972. *The Doomsday Syndrome*. New York: McGraw-Hill.

Mahoney, Michael S. 2002. "Software as Science—Science as Software." In *History of Computing: Software Issues*, edited by Ulf Hashagen, Reinhard Keil-Slawik, and Arthur Norberg. Berlin: Springer.

Mahoney, Michael S. 2005. "The Histories of Computing(s)." *Interdisciplinary Science Reviews* 30: 119–135.

Mahoney, Michael S. 2011. *Histories of Computing*. Edited by Thomas Haigh. Cambridge, MA: Harvard University Press.

Manegold, Karl-Heinz. 1970. *Universität, Technische Hochschule und Industrie. Ein Beitrag zur Emanzipation der Technik im 19. Jahrhundert unter besonderer Berücksichtigung der Bestrebungen Felix Kleins*. Schriften zur Wirtschafts–und Sozialgeschichte. Vol. 16. Berlin, Germany: Duncker & Humblot.

Manegold, Karl-Heinz. 1981. "Der VDI in der Phase der Hochindustrialisierung." In *Technik, Ingenieure und Gesellschaft*, edited by Karl-Heinz Ludwig, 133–166. Duesseldorf, Germany: VDI Verlag.

Marx, Karl. 1973. *Grundrisse. Foundations of the Critique of Political Economy*. London: Penguin.

Mathias, Paul M., Herbert C. Klotz, and John M. Prausnitz. 1991. "Equation-of-State Mixing Rules for Multicomponent Mixtures: The Problem of Invariance." *Fluid Phase Equilibria* 67: 31–44. https://doi.org/10.1016/0378-3812(91)90045-9.

Mauersberger, Klaus. 1980. "Die Herausbildung der Technischen Mechanik und ihr Anteil bei der Verwissenschaftlichung des Maschinenwesens." *Dresdner Beiträge zur Geschichte der Technikwissenschaften* 2: 1–52.

Mauersberger, Klaus, and Friedrich Naumann. 1998. "Die 'Maschinen-Elemente' Carl von Bachs—ein Standardwerk des Maschinenbaus." In *Carl Julius von Bach (1847–1931)*, 155–168. Stuttgart, Germany: Verlag Konrad Wittwer.

Mayer-Schönberger, Viktor, and Kenneth Cukier. 2013. *Big Data: A Revolution That Will Transform How We Live, Work, and Think*. Boston, MA: Eamon Dolan/Houghton Mifflin Harcourt.

Mayo, Deborah G., and Aris Spanos. 2009. *Error and Inference: Recent Exchanges on Experimental Reasoning, Reliability, and the Objectivity and Rationality of Science*. Cambridge: Cambridge University Press.

McDonough, James M. 2007. *Lectures in Computational Fluid Dynamics of Incompressible Flow: Mathematics, Algorithms and Implementations*. Mechanical Engineering Textbook Gallery. Vol. 4. Lexington: University of Kentucky. https://uknowledge.uky.edu/me_textbooks/4.

McGrayne, Sharon Bertsch. 2011. *The Theory That Would Not Die. How Bayes' Rule Cracked the Enigma Code, Hunted down Russian Submarines, and Emerged Triumphant from Two Centuries of Controversy*. New Haven, CT: Yale University Press.

McMullin, Ernan. 1967. "Introduction: Galileo, Man of Science." In *Galileo, Man of Science*, edited by Ernan McMullin, 3–51. New York: Basic Books.

McShane, Edward J., John L. Kelley, and Franklin V. Reno. 1953. *Exterior Ballistics*. Denver, CO: University of Denver Press.

Meadows, Donella H., Dennis L. Meadows, Jorgen Randers, and William W. Behrens. 1972. *The Limits to Growth. A Report for 'The Club of Rome's' Project on the Predicament of Mankind*. New York: Universe Books.

Meadows, Donella H., John Richardson, and Gerhart Bruckmann. 1982. *Groping in the Dark: The First Decade in Global Modelling*. New York: John Wiley and Sons.

Menghini, Marta. 2019. "Precision Mathematics and Approximation Mathematics. The Conceptual and Educational Role of Their Comparison." In *The Legacy of Felix Klein*, edited by Hans-Georg Weigand, William McCallum, Marta Menghini, Michael Neubrand, and Gert Schubring, 181–201. Berlin and Heidelberg: Springer.

Menghini, Marta, and Gert Schubring. 2016. "Preface to the 2016 Edition." In *Felix Klein: Elementary Mathematics from a Higher Standpoint 3: Precision Mathematics and Application Mathematics*, v–x. Berlin and Heidelberg: Springer.

Mesarovic, Mihajlo D., and Eduard C. Pestel. 1974. *Mankind at the Turning Point. Second Report to the Club of Rome*. New York: New American Library.

Metropolis, Nicholas, Jack Howlett, and Gian-Carlo Rota. 1980. *A History of Computing in the Twentieth Century*. New York: Academic Press.

Mindell, David A. 2000. "Automation's Finest Hour. Radar and System Integration in World War II." In *Systems, Experts, and Computers: The Systems Approach in Management and Engineering, World War II and After*, edited by Agatha C. Hughes and Thomas P. Hughes, 27–56. Cambridge, MA: MIT Press.

Mindell, David A. 2002. *Between Human and Machine: Feedback, Control, and Computing Before Cybernetics*. Baltimore, MD: Johns Hopkins University Press.

Mira, Antonietta. 2005. "MCMC Methods to Estimate Bayesian Parametric Models." In *Bayesian Thinking: Modeling and Computation*, edited by Dipak Dey and C. R. Rao, 415–436. Handbook of Statistics. Vol. 25. Amsterdam, Netherlands: Elsevier.

Mirowski, Philip. 2002. *Machine Dreams: Economics Becomes a Cyborg Science*. Cambridge: Cambridge University Press.

Mitchell, Sandra. 2009. *Unsimple Truths. Science, Complexity, and Policy*. Chicago: University of Chicago Press.

Morgan, Mary S. 2012. *The World in the Model. How Economists Work and Think*. Cambridge: Cambridge University Press.

Morgan, Mary S., and Margaret Morrison. 1999. *Models as Mediators. Perspectives on Natural and Social Science*. Cambridge: Cambridge University Press.

Morrison, Margaret. 1999. "Models as Autonomous Agents." In *Models as Mediators. Perspectives on Natural and Social Science*, 38–65. Cambridge: Cambridge University Press.

Morrison, Margaret. 2009. "Models, Measurement, and Computer Simulation. The Changing Face of Experimentation." *Philosophical Studies* 143: 33–57.

Morrison, Margaret. 2015. *Reconstructing Reality. Models, Mathematics, and Simulations*. New York: Oxford University Press.

Mulliken, Robert S. 1989. *Life of a Scientist. An Autobiographical Account of the Development of Molecular Orbit Theory*. Berlin and Heidelberg: Springer.

Mulliken, Robert S., and Clemens C. J. Roothaan. 1959. "Broken Bottlenecks and the Future of Molecular Quantum Mechanics." *Proceedings of the National Academy of Sciences of the United States of America* 45: 394–398.

Münster, Sebastian. 1551. *Rudimenta Mathematica*. Basel, Switzerland: Petri.

Nahin, Paul J. 1988. *Oliver Heaviside: Sage in Solitude. The Life, Work, and Times of an Electrical Genius of the Victorian Age*. New York: IEEE Press.

NAS, National Academy of Sciences—National Research Counsil, Committee on Computers in Chemistry. 1971. *Computational Support for Theoretical Chemistry*. Washington, DC: National Academy of Sciences.

Neal, Radford M. 1998. "Philosophy of Bayesian Inference." https://www.cs.toronto.edu/~radford/res-bayes-ex.html.

Newton, Isaac. 1999. *The Principia: Mathematical Principles of Natural Philosophy*. Berkeley: University of California Press.

November, Joseph A. 2012. *Biomedical Computing: Digitizing Life in the United States*. Baltimore, MD: Johns Hopkins University Press.

November, Joseph A. 2020. "Forsythe, George Elmer." *Complete Dictionary of Scientific Biography*. https://www.encyclopedia.com/science/dictionaries-thesauruses-pictures-and-press-releases/forsythe-george-elmer.

Ntzoufras, Ioannis. 2009. *Bayesian Modeling Using WinBUGS*. New York: John Wiley and Sons.

Nye, Mary Jo. 1993. *From Chemical Philosophy to Theoretical Chemistry*. Berkeley: University of California Press.

Nylander, Johan A. A., James C. Wilgenbusch, Dan L. Warren, and David L. Swofford. 2008. "AWTY (Are We There Yet?): A System for Graphical Exploration of MCMC Convergence in Bayesian Phylogenetics." *Bioinformatics* 24 (4): 581–583.

Oberkampf, William L., and Christopher J. Roy. 2010. *Verification and Validation in Scientific Computing*. Cambridge: Cambridge University Press.

Oreskes, Naomi. 2000. "Why Predict? Historical Perspectives in Prediction in Earth Science." In *Science, Decision Making, and the Future of Nature*, edited by Dan Sarewitz, Roger A. Pielke, Jr., and Radford Byerly, 23–40. Washington, DC: Island Press.

Orrell, David. 2007. *The Future of Everything: The Science of Prediction*. New York: Thunder's Mouth Press.

Otte, Michael. 1989. "Die Auseinandersetzungen zwischen Mathematik und Technik als Problem der historischen Rolle und des Typus von Wissenschaft." In *Mathematik und Technik im 19. Jahrhundert*, edited by Susann Hensel, Karl-Norbert Ihmig, and Michael Otte, 149–214. Goettingen, Germany: Vandenhoeck & Ruprecht.

Otte, Michael. 1993. *Das Formale, das Soziale und das Subjektive*. Frankfurt, Germany: Suhrkamp.

Park, Buhm Soon. 2003. "The 'Hyperbola of Quantum Chemistry': The Changing Practice and Identity of a Scientific Discipline in the Early Years of Electronic Digital Computers, 1945–65." *Annals of Science* 60: 219–247.

Park, Buhm Soon. 2009. "Between Accuracy and Manageability: Computational Imperatives in Quantum Chemistry." *Historical Studies in the Natural Sciences* 39 (1): 32–62.

Parr, Robert G. 1990. "On the Genesis of a Theory." *International Journal of Quantum Chemistry* 37 (4): 327–347.

Parr, Robert G., David P. Craig, and Ian G. Ross. 1950. "Molecular Orbital Calculations of the Lower Excited Electronic Levels of Benzene, Configuration Interaction Included." *Journal of Chemical Physics* 18 (12): 1561–1563.

Parr, Robert G., and Bryce L. Crawford, Jr. 1952. "National Academy of Sciences Conference on Quantum-Mechanical Methods in Valence Theory." *Proceedings of the National Academy of Sciences of the United States of America* 38: 547–553.

Pearl, Judea. 1995. "Causal Diagrams for Empirical Research." *Biometrika* 82 (4): 669–710.

Perdew, John P., Adrienn Ruzsinszky, Jianmin Tao, Viktor N. Staroverov, Gustavo E. Scuseria, and Gábor I. Csonka. 2005. "Prescription for the Design and Selection of Density Functional Approximations: More Constraint Satisfaction with Fewer Fits." *The Journal of Chemical Physics* 123 (6): 62201.

Peyerimhoff, Sigrid D. 2002. "The Development of Computational Chemistry in Germany." In *Reviews in Computational Quantum Chemistry*. Vol. 18, edited by Kenny B. Lipkowitz and Donald B. Boyd, 257–291. New York: VCH Publishers.

Pickering, Andrew. 1995. *The Mangle of Practice. Time, Agency, and Science*. Chicago: University of Chicago Press.

Pietruska, Jamie L. 2017. *Prediction and Uncertainty in Modern America*. Chicago and London, UK: University of Chicago Press.

Pitt, Joseph. 2011. *Doing Philosophy of Technology. Essays in a Pragmatist Spirit*. Dordrecht, Netherlands: Springer.

Platzman, George W. 1967. "A Retrospective View of Richardson's Book on Weather Prediction." *Bulletin of the American Meteorological Society* 48 (8): 514–550.

Pople, John A., and Gerald A. Segal. 1966. "Approximate Self-Consistent Molecular Orbital Theory. III. CNDO Results for AB2 and AB3 Systems." *Journal of Chemical Physics* 44 (9): 3289–3296.

Porter, Theodore N. 1995. *Trust in Numbers. The Pursuit of Objectivity in Science and Public Life*. Princeton, NJ: Princeton University Press.

Puchta, Susann. 1998. "Die Stellung des Ingenieurs und Technikwissenschaftlers Carl Julius von Bach zur Mathematik—ein Beitrag zum Wirken von Bachs bei der Entwicklung der Technikwissenschaften." In *Carl Julius von Bach (1847–1931)*, edited by Friedrich Naumann, 185–208. Stuttgart, Germany: Verlag Konrad Wittwer.

Redmond, Kent C., and Thomas M. Smith. 1980. *Project Whirlwind. The History of a Pioneer Computer*. Bedford, MA: Digital Equipment Corporation.

Redner, Sidney. 2004. "Citation Statistics from More Than a Century of Physical Review." *ArXiv: Physics and Society*. https://arxiv.org/pdf/physics/0407137v1.pdf.

Reuleaux, Franz. 1875. *Lehrbuch der Kinematik*. Vol. 1. Braunschweig, Germany: Vieweg.

Reuleaux, Franz. 1877. *Briefe aus Philadelphia*. Braunschweig, Germany: Vieweg.

Reuleaux, Franz. 1889. *Der Konstrukteur. Ein Handbuch zum Gebrauch beim Maschinen-Entwerfen*. 4th ed. Braunschweig, Germany: Vieweg.

Reuleaux, Franz. 1900. *Lehrbuch der Kinematik*. Vol. 2. Braunschweig, Germany: Vieweg.

Reynolds, Terry. 1991. "The Engineer in Nineteenth-Century America." In *The Engineer in America. A Historical Anthology from Technology and Culture*, edited by Terry Reynolds, 7–26. Chicago: University of Chicago Press.

Richardson, Lewis Fry. 1922. *Weather Prediction by Numerical Process*. Cambridge: Cambridge University Press.

Richardson, Lewis Fry. 1965. *Weather Prediction by Numerical Process. Second Unaltered and Unabridged Edition with New Introduction by Sydney Chapman*. 2nd ed. New York: Dover.

Richenhagen, Gottfried. 1985. *Carl Runge (1856–1927): Von der reinen Mathematik zur Numerik*. Goettingen, Germany: Vandenhoeck & Ruprecht.

Riedler, Alois. 1893. *Amerikanische Technische Lehranstalten. Bericht im Auftrag des Herrn Kultus-Ministers erstattet von A. Riedler*. Berlin: Simion.

Riedler, Alois. 1895. "Zur Frage der Ingenieurerziehung." *Zeitschrift des Vereins Deutscher Ingenieure* 39: 951–959.

Riedler, Alois. 1896. "Die Ziele der Technischen Hochschulen." *Zeitschrift des Vereins Deutscher Ingenieure* 40: 301–309, 337–346, 374–375.

Riedler, Alois. 1898. *Unsere Hochschulen und die Anforderungen des zwanzigsten Jahrhunderts*. Berlin: Seydel.

Riedler, Alois. 1899. "Die Technischen Hochschulen und die wissenschaftliche Forschung (Rektoratsrede TH Charlottenburg)." *Zeitschrift des Vereins Deutscher Ingenieure* 43: 841–844.

Riedler, Alois. 1916. *Emil Rathenau und das Werden der Großwirtschaft*. Berlin: Julius Springer.

Roache, Patrick J. 1982. *Computational Fluid Dynamics*. Albuquerque, NM: Hermosa Publishers.

Robins, Benjamin. 1972. *New Principles of Gunnery*. Richmond, UK: Richmond.

Robinson, Arthur L. 1980. "Plug Pulled on Chemistry Computer Center." *Science* 209 (4464): 1504–1506.

Romeijn, Jan-Willem. 2023. "Philosophy of Statistics." In *The Stanford Encyclopedia of Philosophy*, edited by Edward N. Zalta and Uri Nodelman, Spring 2017 ed. https://plato.stanford.edu/archives/fall2022/entries/statistics/.

Roothaan, Clemens C. J. 1951. "New Developments in Molecular Orbital Theory." *Reviews of Modern Physics* 23: 69–89.

Rosen, William. 2012. *The Most Powerful Idea in the World: A Story of Steam, Industry and Invention.* Chicago: University of Chicago Press.

Rowland, Henry A. 1883. "A Plea for Pure Science." *Science* 2: 242–250.

Rowland, Henry A. 1886. "The Theory of the Dynamo." *Report of the Electrical Conference at Philadelphia in September 1884*, Philadelphia, 86–107.

Runchal, Akshai K. 2009. "Brian Spalding: CFD and Reality—A Personal Recollection." *International Journal of Heat and Mass Transfer* 52 (17–18): 4063–4073.

Russell, Stuart. 2019. *Human Compatible. AI and the Problem of Control.* New York: Penguin Books.

Samaniego, Francisco J., and Dana M. Reneau. 1994. "Toward a Reconciliation of the Bayesian and Frequentist Approaches to Point Estimation." *Journal of the American Statistical Association* 89: 947–957.

Santbech, Daniel. 1561. *Problematum astronomicorum et geometricorum sectiones septem, etc.* Basel, Switzerland: Petri.

Sarewitz, Dan, Roger A. Pielke, Jr., and Radford Byerly, eds. 2000. *Prediction: Science, Decision Making, and the Future of Nature.* Washington, DC: Island Press.

Scerri, Eric R. 1994. "Has Chemistry at Least Been Approximately Reduced to Quantum Mechanics?" *Proceedings of the Biennial Meeting of the Philosophy of Science Association* 1: 160–170.

Scerri, Eric R. 2004. "Just How Ab Initio Is Ab Initio Quantum Chemistry?" *Foundations of Chemistry* 6: 93–116.

Schaefer, Henry F., III. 1986. "Methylene: A Paradigm for Computational Quantum Chemistry." *Science* 231 (4742): 1100–1107.

Schaefer, Henry F., III. 1988. "A History of Ab Initio Computational Chemistry: 1950–1960." *Tetrahedron Computer Methodology* 1 (2): 97–102.

Schappals, Michael, Andreas Mecklenfeld, Leif Kröger, Vitalie Botan, Andreas Köster, Simon Stephan, Edder J. García, et al. 2017. "Round Robin Study: Molecular Simulation of Thermodynamic Properties from Models with Internal Degrees of Freedom." *Journal of Chemical Theory and Computation* 13 (9): 4270–4280. https://doi.org/10.1021/acs.jctc.7b00489.

Schneider, Ascanio, and Armin Masé. 1970. *Railway Accidents of Great Britain and Europe.* Newton Abbot, UK: David and Charles.

Schweber, Sylvan S. 1986. "Shelter Island, Pocono, and Oldstone: The Emergence of American Quantum Electrodynamics after World War II." *Osiris* 2: 265–302.

Schweber, Sylvan S. 1990. "The Young John Clarke Slater and the Development of Quantum Chemistry." *Historical Studies in the Physical and Biological Sciences* 20 (2): 339–406.

Seefried, Elke. 2015. *Zukünfte. Aufstieg und Krise der Zukunftsforschung 1945–1980.* Berlin: de Gruyter.

Seely, Bruce. 1993. "Research, Engineering, and Science in American Engineering Colleges: 1900–1960." *Technology and Culture* 34 (2): 344–386.

Segre, Michael. 1983. "Torricelli's Correspondence on Ballistics." *Annals of Science* 40: 489–499.

Segre, Michael. 1991. *In the Wake of Galileo.* New Brunswick, NJ: Rutgers University Press.

Sengupta, Tapan K., and Srikanth B. Talla. 2004. "Robins–Magnus Effect: A Continuing Saga." *Current Science* 86 (7): 1033–36.

Shapin, Steven, and Simon Shaffer. 1985. *Leviathan and the Air-Pump: Hobbes, Boyle, and the Experimental Life.* Princeton, NJ: Princeton University Press.

Sherden, William A. 1997. *The Fortune Sellers. The Big Business of Buying and Selling Predictions.* New York: John Wiley and Sons.

Shinn, Terry. 1980. "From 'Corps' to 'Profession'; the Emergence and Definition of Industrial Engineering in Modern France." In *The Organization of Science and Technology in France, 1808–1914*, edited by Robert Fox and George Weisz, 183–203. Cambridge: Cambridge University Press.

Siacci, Francesco. 1892. *Balistique extérieure.* Paris: Berger-Levrault.

Siegel, Eric. 2013. *Predictive Analytics: The Power to Predict Who Will Click, Buy, Lie, or Die.* New York: John Wiley and Sons.

Silver, Nate. 2012. *The Signal and the Noise. Why So Many Predictions Fail—But Some Don't.* New York: Penguin Press.

Simmons, Harvey. 1973. "System Dynamics and Technocracy." In *Models of Doom—A Critique of The Limits to Growth*, edited by H. S. D. Cole, Christopher Freeman, Marie Jahoda, and K. L. R. Pavitt, 192–208. New York: Universe Books.

Simões, Ana I. 2002. "Dirac's Claim and the Chemists." *Physics in Perspective* 4: 253–266.

Simões, Ana I. . 2003. "Chemical Physics and Quantum Chemistry in the Twentieth Century." In *The Cambridge History of Science, Vol. 5: The Modern Physical and Mathematical Sciences*, edited by Mary Jo Nye, 394–412. Cambridge: Cambridge University Press.

Simon, Herbert A. 1996. *Models of My Life.* Cambridge, MA: MIT Press.

Smith, Adrian F. M. 1984. "Present Position and Potential Developments. Some Personal Views: Bayesian Statistics." *Journal of the Royal Statistical Society A* 147 (2): 245–259.

Smith, Adrian F. M. 1988. "What Should Be Bayesian about Bayesian Software?" In *Bayesian Statistics*. Vol. 3, edited by José M. Bernardo, Morris. H. DeGroot, David V. Lindley, and Adrian F. M. Smith, 429–435. Oxford: Oxford University Press.

Smith, Adrian F. M., and Gareth O. Roberts. 1993. "Bayesian Computation via the Gibbs Sampler and Related Markov Chain Monte Carlo Methods (with Discus)." *Journal of the Royal Statistical Society B* 55 (1): 3–23.

Smith, Brian Cantwell. 2019. *The Promise of Artificial Intelligence. Reckoning and Judgement*. Cambridge, MA: MIT Press.

Smith, George E. 2002. "From the Phenomenon of the Ellipse to an Inverse-Square Force: Why Not?" In *Reading Natural Philosophy: Essays in the History and Philosophy of Science and Mathematics*, edited by David B. Malament, 31–70. Chicago: Open Court.

Span, Roland, Eric W. Lemmon, Richard T. Jacobsen, Wolfgang Wagner, and Akimichi Yokozeki. 2000. "A Reference Equation of State for the Thermodynamic Properties of Nitrogen for Temperatures from 63.151 to 1000 K and Pressures to 2200 MPa." *Journal of Physical and Chemical Reference Data* 29: 1361–1433. https://doi.org/10.1063/1.1349047.

Stäckel, Paul. 1915. *Die mathematische ausbildung der architekten, chemiker und ingenieure an den Deutschen technischen hochschulen*. Leipzig, Germany: Teubner.

Steele, Brett D. 1994. "Muskets and Pendulums: Benjamin Robbins, Leonhard Euler, and the Ballistics Revolution." *Technology and Culture* 35: 348–382.

Steele, Brett D., and Tamara Dorland. 2005. *The Heirs of Archimedes: Science and the Art of War through the Age of Enlightenment*. Cambridge, MA: MIT Press.

Stern, Hal. 2011. "Discussion of 'Statistical Inference: The Big Picture' by R. E. Kass." *Statistical Science* 26 (1): 17–18.

Stodola, Aurel. 1897. "Die Beziehungen der Technik zur Mathematik (Vortrag erster Internationaler Mathematikerkongress 1897 Zürich)." *Zeitschrift des Vereins Deutscher Ingenieure* 41: 1257–1260.

Strong, Edward W. 1936. *Procedures and Metaphysics. A Study in the Philosophy of Mathematical-Physical Science in the Sixteenth and Seventeenth Centuries*. Berkeley: University of California Press.

Sumpner, William E. 1928. "Heaviside's Fractional Differentiator." *Proceedings of the Physical Society* 41: 404–425.

Swain, Geo F. 1893. "Comparison between American and European Methods in Engineering Education." *SPEE Proceedings* 1: 75–102.

Tal, Eran. 2013. "Old and New Problems in Philosophy of Measurement." *Philosophy Compass* 8 (12): 1159–1173.

Talbott, William. 2016. "Bayesian Epistemology." In *The Stanford Encyclopedia of Philosophy*, edited by Edward N. Zalta. https://plato.stanford.edu/archives/win2016/entries/epistemology-bayesian/.

Taleb, Nassim N. 2007. *The Black Swan: The Impact of the Highly Improbable*. New York: Random House.

Tetmajer, Ludwig. 1901. *Die gesetze der knickungs–Und der zusammengesetzten Druckfestigkeit der technisch wichtigsten Baustoffe*. Leipzig, Germany and Wien: Franz Deuticke.

Thomas, A., David J. Spiegelhalter, and Walter R. Gilks. 1992. "BUGS: A Program to Perform Bayesian Inference Using Gibbs Sampling." In *Bayesian Statistics 4*, edited by José M. Bernardo, James O. Berger, A. P. David, and Adrian F. M. Smith, 837–842. Oxford: Oxford University Press.

Thurston, Robert H. 1884. "The Mission of Science." *Proceedings of the American Association of Science, 33rd Meeting*, Philadelphia, 227–253.

Thurston, Robert H. 1893a. "Technical Education in the United States: Its Social, Industrial, and Economic Relations to Our Progress." *Transaction of the American Society of Mechanical Engineers* 14: 855–877, 950–955.

Thurston, Robert H. 1893b. "Technological Schools: Their Purpose and Its Accomplishment." *Scientific American Supplement* 934: 14920–14921, 14940–14941.

Thurston, Robert H. 1893c. "The Equipment of Engineering Schools." *Scientific American Supplement* 939: 15008–15009.

Thurston, Robert H. 1896. "The Engineering Experiment Station of Sibley College, at Cornell University." *The Sibley Journal of Engineering* 10 (7): 269–287.

Titterington, Michael D. 2004. "Statistical Modeling and Computation." In *Applied Bayesian Modeling and Causal Inference from Incomplete-Data Perspectives*, edited by Andrew Gelman and Xiao-Li Meng, 189–194. New York: John Wiley and Sons.

Toffler, Alvin, ed. 1972. *The Futurists*. New York: Random House.

Torricelli, Evangelista. 1644. "Opera Geometrica." http://echo.mpiwg-berlin.mpg.de/MPIWG:WHZEF9W9.

Truesdell, Clifford. 1954. "Rational Fluid Mechanics, 1657–1765." *Opera Omnia, Series 2* 12: 19–26.

Truesdell, Clifford. 1968. "A Program toward Rediscovering the Rational Mechanics of the Age of Reason." In *Essays in the History of Mechanics*, 85–137. New York: Springer.

Truesdell, Clifford. 1984. *An Idiot's Fugitive Essays on Science*. New York: Springer.

Turner, Fred. 2006. *From Counterculture to Cyberculture. Stewart Brand, the Whole Earth Network, and the Rise of Digital Utopianism.* Chicago: University of Chicago Press.

Valderrama, José O. 2003. "The State of the Cubic Equation of State." *Industrial Engineering Chemistry Research* 42 (7): 1603–1618.

Valleriani, Matteo. 2010. *Galileo Engineer.* Dordrecht, Netherlands: Springer.

Valleriani, Matteo. 2013. *Metallurgy, Ballistics and Epistemic Instruments: The Nova Scientia of Niccolò Tartaglia, A New Edition.* Edition Open Access. Berlin, Germany: Max Planck Research Library for the History and Development of Knowledge. https://edition-open-sources.org/sources/6/index.html.

Vincenti, Walter G. 1990. *What Engineers Know and How They Know It.* Baltimore, MD: Johns Hopkins University Press.

Voskuhl, Adelheid. 2016. "Engineering Philosophy: Theories of Technology, German Idealism, and Social Order in High-Industrial Germany." *Technology and Culture* 57 (4): 721–752.

Waldrop, M. Mitchel. 2001. *The Dream Machine. J.C.R. Licklider and the Revolution That Made Computing Personal.* New York: Penguin Books.

Walsh, Lynda. 2013. *Scientists as Prophets. A Rhetorical Genealogy.* New York: Oxford University Press.

Warwick, Andrew R. 1992. "Cambridge Mathematics and Cavendish Physics: Cunningham, Campbell, and Einstein's Relativity, 1905–1911. Part 1, The Uses of Theory." *Studies in History and Philosophy of Science* 23: 625–656.

Warwick, Andrew R. 1995. "The Laboratory of Theory or What's Exact about the Exact Sciences?" In *The Values of Precision,* edited by M. Norton Wise, 311–351. Princeton, NJ: Princeton University Press.

Wei, Ya Song, and Richard J. Sadus. 2000. "Equations of State for the Calculation of Fluid-Phase Equilibria." *Aiche Journal* 46 (1): 169–196.

Weisbach, Julius F. 1880. *Lehrbuch der ingenieur- und maschinenmechanik.* Vol. 2 of *Die ingenieur- und arbeitsmaschinen.* Braunschweig, Germany: Vieweg.

Weisberg, Michael. 2013. *Simulation and Similarity. Using Models to Understand the World.* New York: Oxford University Press.

Westfall, Richard S. 1985. "Scientific Patronage: Galileo and the Telescope." *Isis* 76 (1): 11–30.

Wieber, Frédéric, and Alexandre Hocquet. 2020. "Models, Parameterizations, and Software: Epistemic Opacity in Computational Science." *Perspectives on Science* 28 (5): 610–629.

Wiener, Martin J. 1981. *English Culture and the Decline of the Industrial Spirit, 1850–1980*. Cambridge: Cambridge University Press.

Wiener, Norbert. 1948. *Cybernetics: Or Control and Communication in the Animal and the Machine*. Cambridge, MA: Technology Press and John Wiley.

Williamson, Jon. 2005. *Bayesian Nets and Causality*. Oxford: Oxford University Press.

Wilson, Mark. 2006. *Wandering Significance. An Essay on Conceptual Behavior*. New York: Oxford University Press.

Wilson, Mark. 2017. *Physics Avoidance. Essays in Conceptual Strategy*. New York: Oxford University Press.

Winsberg, Eric. 1999. "Sanctioning Models: The Epistemology of Simulation." *Science in Context* 12 (2): 275–292.

Winsberg, Eric. 2003. "Simulated Experiments: Methodology for a Virtual World." *Philosophy of Science* 70: 105–125.

Winsberg, Eric. 2010. *Science in the Age of Computer Simulation*. Chicago: University of Chicago Press.

Wise, M. Norton. 2004. *Growing Explanations: Historical Perspectives on Recent Science*. Durham, UK: Duke University Press.

Woody, Andrea I. 2013. "How Is the Ideal Gas Law Explanatory?" *Science and Education* 22: 1563–1580.

Worrall, John. 2010. "Theory Confirmation and Novel Evidence. Error, Tests, and Theory Confirmation." In *Error and Inference: Recent Exchanges on Experimental Reasoning, Reliability, and the Objectivity and Rationality of Science*, edited by Deborah G. Mayo and Aris Spanos, 125–154. Cambridge: Cambridge University Press.

Yavetz, Ido. 1995. *From Obscurity to Enigma. The Work of Oliver Heaviside, 1872–1889*. Basel: Birkhaeuser Verlag.

Zellner. 1988. "A Bayesian Era." In *Bayesian Statistics*, Vol. 3, edited by José M. Bernardo, Morris. H. DeGroot, David V. Lindley, and Adrian F. M. Smith, 509–516. Oxford: Oxford University Press.

Zielinski, Jan. 1995. *Ludwig von Tetmajer Przerwa 1850–1905*. Meilen, Switzerland: Verein für wirtschaftshistorische Studien.

Index